APPLICATION SPECIFIC INTEGRATED CIRCUITS - TECHNOLOGIES, DIGITAL SYSTEMS AND DESIGN METHODOLOGIES

Edited by **Edward Fisher**

Application Specific Integrated Circuits - Technologies, Digital Systems and Design Methodologies
http://dx.doi.org/10.5772/intechopen.73989
Edited by Edward Fisher

Contributors

Erik Den Hartigh, Claire Stolwijk, Roland Ortt, Wim Vanhaverbeke, Arya Wicaksana, Dareen Kusuma Halim, Dicky Hartono, Felix Lokananta, Sze-Wei Lee, Mow-Song Ng, Chong-Ming Tang, Carmine Pappalettere, Caterina Casavola, Luciano Lamberti, Vincenzo Moramarco, Giovanni Pappalettera, Muhammad Athar Javed Sethi, Momil Ijaz, Huma Urooj, Fawnizu Azmadi Hussin, Edward M. D. Fisher

Notice

Statements and opinions expressed in the chapters are these of the individual contributors and not necessarily those of the editors or publisher. No responsibility is accepted for the accuracy of information contained in the published chapters. The publisher assumes no responsibility for any damage or injury to persons or property arising out of the use of any materials, instructions, methods or ideas contained in the book.

First published in London, United Kingdom, 2019 by IntechOpen
IntechOpen is the global imprint of INTECHOPEN LIMITED, registered in England and Wales, registration number: 11086078, The Shard, 25th floor, 32 London Bridge Street
London, SE19SG – United Kingdom
Printed in Croatia

British Library Cataloguing-in-Publication Data
A catalogue record for this book is available from the British Library

Additional hard copies can be obtained from orders@intechopen.com

Application Specific Integrated Circuits - Technologies, Digital Systems and Design Methodologies, Edited by Edward Fisher
p. cm.
Print ISBN 978-1-78985-847-1
Online ISBN 978-1-78985-848-8

We are IntechOpen,
the world's leading publisher of
Open Access books
Built by scientists, for scientists

4,100+
Open access books available

116,000+
International authors and editors

120M+
Downloads

Our authors are among the

151
Countries delivered to

Top 1%
most cited scientists

12.2%
Contributors from top 500 universities

CLARIVATE ANALYTICS
BOOK
CITATION
INDEX
INDEXED

WEB OF SCIENCE™

Selection of our books indexed in the Book Citation Index
in Web of Science™ Core Collection (BKCI)

Interested in publishing with us?
Contact book.department@intechopen.com

Numbers displayed above are based on latest data collected.
For more information visit www.intechopen.com

Meet the editor

Edward M.D. Fisher (IEEE Member 2008) received his MEng degree in Electronic and Electrical Engineering from the University of Edinburgh in 2009, interning at STMicroelectronics researching automatic exposure algorithms. After completing his PhD in single-photon avalanche diode (SPAD) arrays in CMOS for optical communications (University of Edinburgh, 2015), he began work on high-speed parallel data acquisition systems as a member of the Agile Tomography Group, University of Edinburgh. This work aimed to provide chemical species tomography diagnostics for aero-engines in collaboration with Rolls-Royce, Royal Dutch Shell, and academia within the FLITES project (www.flites.eu). He also previously worked with a bioelectrical impedance spectroscopy company utilizing microfluidics for label-less cell counting, classification, and sorting via electrophoresis. He periodically revisited SPADs with a 2017 chapter on ASIC readout topologies for high-speed data communications and a 2018 book chapter on a robust literature review of the historical development of avalanche gain devices and solid-state photon counting (1900–1969). He is now with Coda Octopus Products Ltd., working on digital design and data acquisition for deep-sea sonar. His responsibilities include adaptions to the sonar pulse waveform for military customers and the implementation of new bottom/object detection algorithms based on the Canny filter. He has an interest in digital signal processing for low-noise data acquisition and methods in which ASIC and FPGA design can be taught.

Contents

Preface

To Rebecca Price. For always standing by my side.

Since the 1970s, the field of integrated circuits (ICs) has matured rapidly and has made significant contributions to our society. Each day we hear of new photolithography methods able to produce novel nanometre-scale structures that can become new transistor devices, and new systems produced with the latest complementary metal-oxide semiconductor (CMOS) process node. We are often reminded of the history of application-specific integrated circuits (ASICs) through discussions of Moore's law—the rough doubling of transistor counts every 18 months—and of course through discussions of where the IC field will go as Moore's law starts to slow, saturate, and become uneconomical for traditional planar (two-dimensional) ICs.

There are now a wide variety of subfields under the ASIC umbrella. Each of these has received significant academic attention, commercial financing, and industrial research and development for large-scale manufacturing during different phases of their maturation process. We can chiefly split the ASIC field into three substrata. The first deals with digital ASICs such as central processing units, graphics processing units, microcontroller units, and of course memory ICs and custom-made digital signal processing ICs. The second substratum covers analog functions such as analog-to-digital converters (ADCs), digital-to-analog converters (DACs), oscillators such as phase-locked loops, amplifiers and sensor interfaces or native semiconductor sensors such as photodiode detectors, and Hall effect magnetic sensors. The third substratum would be best described as more-than-Moore technologies. These are not an increase in transistor count above Moore's law, but are instead an increase in functionality over and above the existing complex analog or digital functionality. For example, a fully integrated device bringing together low-noise high-accuracy analog circuits, microelectromechanical systems, optical sensing, and significant volumes of high-speed digital-signal processing offers a far greater functionality per unit volume than each of these technologies operating in a nonintegrated or separate IC manner.

This book brings together a small collection of chapters covering specific aspects of the ASIC field. The first section, or rather the introductory chapter (Chapter 1), discusses the field in general terms, providing the keen student with key literature and textbooks for a variety of subfields within ASIC design. The combination of sources in this chapter's reference list is a superb overview and introduction to the design methods used in the field, and sources have been selected to provide those interested with a starting point for a selection of disparate specialisms. The ASIC field and the literature are somewhat of a rabbit-hole, with these key sources being suitable jumping points for more detailed reading.

Section 2 covers aspects of the ASIC field itself in terms of business strategy and design practices. In Chapter 2, an assessment is made of how the market and company finances work in the field in the face of a base requirement of a diverse ASIC technology portfolio and the increasing barriers to entry into this area of modern technology.

In Section 3, a few critical aspects of ASIC design are presented. In Chapter 3, a case study is used to provide some lessons learnt that can aid in successful first-time ASIC designs taken to tape-out. A number of topics for the design of large, complex system on chip (SoC) designs are mirrored between this chapter and the introduction (Chapter 1). In Chapter 4, the author has chosen to investigate variants of the packet-switching protocols of the modern network on chip. These are becoming crucial parts of multiprocessor SoC architectures because arrays of individually controllable processing elements can be efficiently allocated for the incoming data and instructions, and the packet routing strategy can afford a significant fault tolerance advantage over traditional bus architectures.

Finally, in Section 4, Chapter 5 provides an optical inspection method allowing fault detection and diagnostics to be performed on packaged and system-integrated ASICs. While the example uses a regulator IC within a basic small-outline transistor package, the principle of using optical methods to investigate thermal and mechanical stresses upon the silicon could provide crucial investigative information to aid system, ASIC, and design failure mechanisms.

I would like to express my gratitude to the publishers (InTechOpen) for their help with this book and the authors of accepted manuscripts for their work and patience. For the interested readers, I would like to suggest they keep an eye on both the *IEEE Journal of Solid-State Circuits* and the annual conferences: the San Francisco-based International Solid-State Circuits Conference, the European Solid-State Circuits Conference, and the Asian Solid-State Circuits Conference. Together with numerous books within the ASIC field, these track novelty within the field covering topics such as new Intel/ARM processors, Nvidia GPU structures, cutting-edge ADC and DAC technologies, and the iterative progress being made in CMOS and Bi-CMOS nanometre-scale technologies.

Dr. Edward M. D. Fisher
Coda Octopus Products Ltd.
Scotland, United Kingdom

Introduction

Chapter 1

Introductory Chapter: ASIC Technologies and Design Techniques

Edward M.D. Fisher

Additional information is available at the end of the chapter

http://dx.doi.org/10.5772/intechopen.84416

1. Introduction

This chapter aims to introduce the field of application specific integrated circuits (ASICs) and provide a basic reference list of crucial design techniques for the interested student. Modern designs using complementary metal oxide semiconductor (CMOS) technologies are at the forefront of nanoscale mass fabrication with recent CMOS processing nodes pushing towards 7 nm feature sizes. While the fabrication of structures on Silicon, Germanium, Gallium Arsenide and other substrates has advanced, so too have the systems we create [1, 2]. The rapidly increasing complexity inherent in these systems – be they digital or analogue in nature – offers its own challenges for engineers. This book therefore offers a small insight into this exciting field.

Modern silicon design plays a key part in our global, highly interconnected economy and has truly allowed society to advance on several fronts.

- We now have sensors and interfaces to the real world that digitise high-speed analog signals into easily manipulated digital signals at staggeringly high sample rates, well into the GHz range [3]. Modern analog-to-digital converters (ADCs) when combined with precision, low-noise operational amplifiers (op-amps) can accurately measure nano-volt amplitude signals and can separate signals from interfering background noise using a variety of signal processing techniques. The field's prowess in analog ASICs is well exemplified by sensors and actuators that can interface directly with human brain tissue.

- Custom and general use microprocessors [4] have been a shining light of progress on complex integrated digital systems and now offer us continued growth for our increasingly complex computational needs. The improvements in super-computing are a testament to our ongoing, field-wide interest in increasing the floating-point operations per second, while recent work aims to tackle the energy required per computation. By combining high-performance, low-power ASICs and novel computational architectures (systolic arrays, array processing, graphics processing units, etc.), super-computers and data-centres are now able to provide enough resource for a variety of high-dimensional simulation problems. So powerful are modern central processing units (CPUs) that even embedded processors can run highly-nested (i.e. deep) neural networks of many thousands of complex artificial neurons, which has directly enabled the rise of machine learning (ML), deep learning (DL) and artificial intelligence (AI). Likewise, that same computation with low-energy and high-integration progress allows many of us to have a computer within our pocket that outstrips the military and commercial computers of the 80s.

- Integrated, solid-state memory circuits [5] continue to become smaller, of higher access speed, of higher long-term reliability and of lower power, now replacing traditional hard disk drives in many front-line computing tasks. This has allowed society to store vast quantities of data and has lead – with ease of access and computation – to the rise of big data analytics and an increased interest in statistical and data-guided adaptive signal processing.

- Rapid strides in digital image sensor designs [6] now allow individual pixels of less than 900 nm in width (very close to the diffraction limit), arrays of hundreds of megapixels, sensors able to be used in 645 medium format camera systems and even the counting and timing of single-photons with sub-10 pico-second timing resolutions. Digital image sensors using various semiconductors, substrates and readout methods can image at incredibly high frame rates – well into the millions of frames/s – and over a very large portion of the electromagnetic spectrum. Telescopes such as Hubble and James Webb use these image sensor ASICs to investigate the cosmos, expanding our scientific knowledge, detecting planets orbiting distant stars and showing the innate artistry of nature. And – if this wasn't enough – digital image sensors for machine vision can be coupled with complex digital signal processing ASICs and CPUs/GPUs running neural networks to provide automatic pedestrian detection and avoidance for self-driving cars.

But the progress we have observed in each of the above sub-fields from the 1970s to present has required many difficult issues to be tackled and presents a highly interesting and challenging career for those that are interested in the design details and techniques that are required. In this chapter, Section 2 breaks the ASIC arena down into a set of sub-fields providing some key texts for each. While a little more emphasis is placed on digital circuits and CMOS optical detectors and image sensors, this is a consequence of this author's interests. To this end, Section 3 discusses some of the digital ASIC and programmable logic device hardware description languages (HDLs) that are now routinely used to develop complex systems. Finally, Section 4 discusses three key industry standards – *ISO-9001*, *ISO-26262* and *DO-254* – which emphasise the robustness and design formalism required in the industry. We also discuss the universal design methodology (UDM), which alongside modern ASIC verification standards such as the open verification methodology (OVM) and the universal verification methodology (UVM), introduce some of the key ways in which designers of ASICs can handle increasing design complexity.

2. Suggested references and reading

In this section, a wide selection of texts is references that provide significant context to the design issues inherent within ASIC technologies. The introductory text by Huber [7] provides an insight into how systems were designed during the 1990s. While this does not cover the detail of theory required for design, it introduces us to the manner, methods and tools used within the field. Generally, the below set of resources is split into the overall themes of the field: (i) planar silicon processes and solid-state physics, (ii) analogue transistors and systems, (iii) digital circuit design, and finally, (iv) sensors and interfaces to the real macro world in which we utilise these devices.

2.1. Planar silicon processes and solid-state physics

A crucial text for understanding the nature of electronics using nano-fabrication and doping of semiconductors is the work of Simon Sze [8]. This text not only discusses the way designs can be fabricated in a planar fashion using crystals of semiconductors such as Silicon, but it also introduces many of the microelectronic device structures. For example, bipolar junction transistors (BJTs) and metal-oxide-semiconductor field-effect transistors (MOSFETs) are discussed in detail. Sze undertook much of his research at Bell Telephone laboratories working alongside Walter Brattain, John Bardeen and William Shockley (the inventors of solid-state BJT and MOS transistors). His texts are crucial reading for those interested in the way electrons and holes within semiconductors can be used for analog amplification, digital switching and how semiconductors can be used as sensors.

2.2. Analog transistors, systems, sensors and interfaces

The development of integrated circuits, in particular ASICs and the field of very large-scale integration (VLSI) arguably started with the design and integration of analog circuitry within planar semiconductor manufacturing processes. The theory behind analog systems is well covered by the industry standard texts of Allen and Holberg [9] and Weste and Eshraghian [10]. While for the most part analog designs operate in terms of voltage amplitudes and the amplification, addition, subtraction, filtering and manipulation thereof, it should be noted that a current-mode approach can also be taken [11]. For the case of data communications using currents rather than voltages, authors such as Yuan [12] note that current-mode designs have advantages over their voltage-mode counterparts. As silicon also responds to light, CMOS analogue circuits have found application for many optical sensing methods [13] and of course image sensors [14]. Likewise, many other quantities such as temperature, magnetic fields (via the Hall effect) or ion-concentrations can be measured using CMOS technologies.

2.3. Digital circuit design theory

While modern ASICs – in an effort to increase the levels of system integration – often combine analog and digital circuitry, there has been significant emphasis on digital systems. Of importance for digital ASIC design, the texts by Kaeslin [15], Wakerly [16], and Weste and

Harris [17] can be considered worthwhile purchases for students. Of note, Kaeslin's book [15] covers much of the setup, hold and propagation timing issues that need to be considered during the design of circuits near the upper end of the possible clock speeds for a particular CMOS process node.

Much of the push within the digital ASIC sector has been the rapid development of high-speed digital signal processing (DSP), memory and of course microcontroller and computer architectures. Students interested in the field of DSP are suggested to read the introductory text by Lynn and Fuerst [18]. Likewise, many of the concepts discussed in the context of field-programmable gate arrays (FPGAs) by Hauck and De Hon [19], and Mayer-Baese [20], can be transferred into the digital ASIC domain. The advantage, of course, being that FPGAs offer a method of prototyping or deploying a digital circuit without the high costs associated with CMOS ASIC designs [21]. While computer architecture quickly takes steps away from the low-level details of digital computation, storage and manipulation, the de-facto references for the field are the books by Hennessy and Patterson [22]. The crucial point here is that incredibly complex systems can be created by the prudent use of modular digital designs and a suitable hierarchical design and abstraction strategy.

The languages and tools typical of the digital ASIC design industry are discussed in detail in Section 3. Suffice to say that the hardware description languages (HDLs) of Verilog [23] and VHDL [24] are crucial additions to the digital ASIC designer's repertoire of designs tools.

3. Suggested hardware description languages

While the ASICs of the past were typically drawn by hand – a highly laborious task with a high-risk factor – modern ASICs are almost exclusively developed using complex computer aided design (CAD) packages. The advantages of this are clear, first, that designs can become significantly more complex in a scalable, well maintainable and modular manner and secondly that transient, temporal, thermal, frequency domain, power, parasitic circuit elements and complex second and third order effects can be added in a manner that would not be possible from a hand calculation perspective. CAD packages allow designs to be verified against multiple specifications. Likewise, simulations based upon reliable fitted models can be used to provide such a wide range of correct and erroneous stimuli that we can obtain 99.9% test coverage and a very high expectation of full functionality prior to silicon prototyping. It is possible to run a design against multiple design corners such as low/high power, low/high temperature and fast/slow transistors, allowing simulation to capture a high proportion of likely external factors and combinations thereof.

For analog, MEMs and photonic ICs, the user typically defines the lengths and widths of transistors, silicon features (resonant beams, masses, etc.) or waveguide structures using established theoretical techniques and formulae, and then iterates around this design as the tools iteratively increase the complexities added to the base models.

Digital design now exclusively uses hardware description languages. These were initially used to describe and model digital circuits, however modern ASIC design houses use a HDL to silicon synthesis process whereby the language describes a set of logical gates and sequential (clocked) registers that are synthesised – often via complex optimisation processes – into physical logic gates, registers and interconnections that are provided as standard cells by the CMOS foundry. This type of design entry is often called the register transfer level (RTL). The standard cells are themselves designed using the analog design flows and complex models to provide known digital performance (setup, hold and propagation times, etc.). There are two crucial HDLs used within the field. The first is Verilog [23], while the second is a very high speed integrated circuit (VHSIC) HDL, called VHDL [24]. Typically, Verilog is used for hardware descriptions that are synthesised to Silicon, while VHDL is typically synthesised to programmable devices such as FPGAs [21] and complex programmable logic devices (CPLDs). This split is however a rather grey area with a great deal of company preference. These description languages should therefore be treated as complementary.

While Verilog and VHDL have structures that comprehensively cover design concepts and the required complexity, languages such as SystemVerilog [25] and SystemC are used to provide functional verification of digital logic designs. The ASIC field would therefore write a SystemVerilog testbench that simulates a wide variety of stimuli for a digital module written in Verilog or VHDL. By doing so, we can be assured of the block's functional or logical design before we progress to more complex timing verification steps. As designs have become more complex, so have the testbenches upon which we assess a design's verifiable functionality and use case or test coverage. To handle such complexity, the digital ASIC field has shifted to a higher-level abstraction. It is here we have introduced the universal verification methodology (UVM) [26] which treats the verification of a block using a transaction-based system. The block or design-under-test (DUT) is provided with transactions that are monitored using a secondary process, with the DUT then being given a score based upon the successful completion of that transaction. For example, a two-input 16-bit adder could be given two series of random values. These would be passed through the DUT, while the monitor would compare the DUT output values with the deterministic, design-independent addition.

4. Standards and the universal design methodology (UDM)

As part of any introduction to design methods, practices and theory, we as designers must also pay attention to industry standards. While this is perhaps not given sufficient emphasis within academia, certain aspects of commercial engineering practice need to be used independent of a student's career goals. We should emphasise that students should read up on industry standards such as *ISO9001, ISO-26262* or *DO-254* as these will be used by the clear majority of technology companies and do in fact benefit academic projects.

4.1. Standards

Control of a design project allows us to manage both complexity and risk, while giving customers (or the generalised concept of a project's stakeholder) an assurance as to the robustness of a design. ISO-9001 *"Quality Management Systems - Requirements"* [27] defines a process model whereby tasks are suitably defined, documented and allocated based upon the requirements of a design. The top-down flow of requirements – in complement to a top-down flow of documentation – allows the project to be traceable in terms of decisions taken and/or risk management. Likewise, it also allows design stages to be signed off after they have been demonstrated to capture all requirements and provide verification as to a design's fitness for purpose. A theme within such project management methods is the use of abstraction levels – a concept in common with electrical system design.

- During the design process, it may be necessary to modify or feedback into the previously agreed requirements of a block. While project managers may not need to know the bit-level detail as to the change, it would be crucial to discuss the changes to a project's costs or timeline to ensure that all involved in the project are clear of the implications.

- ISO-9001 [27] is routinely used throughout engineering to ensure clear communication and the management of expectations of all involved within a design. In industry, this would be hardware engineers, project managers, product marketing executives, the board of directors and of course the assurance for the end-user that their problem has been provided with a suitable solution.

For complex electrical systems, the application – for example, military/aviation, automotive or biomedical – often includes end-use standards that impact the bit-level or electrical system design process. As an example, ISO-26262 "Road Vehicles – Functional Safety" [28] defines – in a higher-level manner – the levels of fault and error tolerance that are required when engineers design safety critical systems. In effect, such standards seek to verify and provide assurance that no harm can be caused through poor design or to ensure that under random external fault-causes that the system behaves in a defined and predicable manner.

- One well known implication of this standard is that of redundancy within a system. For example, a set of three redundant control blocks may be created with a voting system passing the control signal to an actuator. A fault in any of the three caused by a timing or power glitch is prevented from causing erroneous action.

- Likewise, a processor responsible for the car's breaks should default to a passenger-safe state or shutdown sequence if a fault is detected within its registers via the use of bit-level parity.

- When data is transmitted – but has implications for safety – error correction codes can be used to ensure that within 99% of cases the message is interpreted correctly at the receiver. The same method also allows compliance by providing a default or shut down for the remaining cases where a message cannot be guaranteed.

DO-254 "Design Assurance Guidance for Airborne Electronic Hardware" [29] defines a further, stringent and often mandatory set of design directives for what it states as complex electronic systems. These are defined to be PLDs, FPGAs and ASICs. The standard sets out five (A to E) levels of compliance each related to a risk level, where the assurance is provided by verification and validation steps that fundamentally demonstrate that the quantified risk factors have been suitably mitigated and handled within the design itself. Depending on the ASIC product, these – alongside other such standards – must be met.

4.2. Universal design methodology

While the universal design methodology (UDM) [30] is often applied to FPGA-based hardware design, it is applicable to a wide range of electronics projects including both analog and digital ASIC designs. The power of the UDM model is its ability to fit well with the existing structures and processes of ISO-9001, while also enforcing a robust formalism in the design process. The pathway within UDM, shown in **Figure 1**, includes multiple opportunities for the design to be specified and constrained in a top down manner, while also providing suitable feedback loops to ensure designs meet all specifications without undue over-engineering. UDM also includes multiple signing off and review processes, each with the purpose of finalising, verifying and capturing issues as early as possible as the design moves towards implementation and shipping. In this way, once the design is passed to physical (gate-level) implementation, i.e. synthesis to Silicon or even custom analog layout, designers should never encounter a new requirement to add a separate reset line, a new state within an initialisation or control state machine or modify the block's frequency domain specification.

UDM contains a significant emphasis on design verification at all stages. For example, the functional simulation of a HDL coded design is used with the design specification to ensure the design both captures all design intent within the specifications and is verifiably functionally correct and in agreement with the agreed (via review) list of specifications. Likewise, once the HDL design is synthesised to Silicon, it is formally verified via gate-level simulation or emulation on a programmable logic platform, to ensure that it meets all timing requirements and that no issues are present using a constrained random approach to testing. Part of the verification task is to demonstrate functional equivalence between the design simulated, verified and agreed upon in previous stages with the computer optimised design that was synthesised. The final review within UDM is crucial as it is at this point that all involved in the design, from marketing and product management right down to hardware and software engineers sign-off the design and prepare for manufacture and final testing of the product. UDM can therefore be instrumental in achieving agreement between professionals of highly

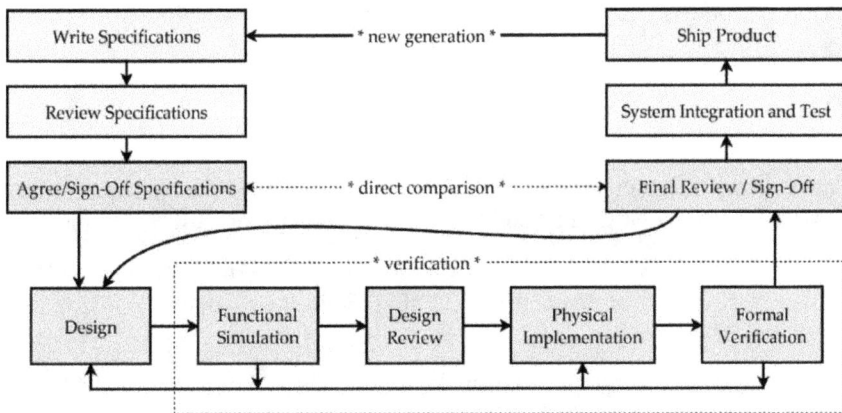

Figure 1. The universal design methodology (UDM), with an emphasis on multiple sign-off processes within the design flow and functional and final design verification against agreed specifications.

disparate backgrounds and provides a mechanism whereby a bit-level hardware design has a proven track record and verifiable performance against a set of high-level requirements.

It should be noted that ASIC design tool vendors such as Cadence and Synopsis include many sign-off checks that complement the sign-off and review steps in UDM. For example, design rule checks (DRC) and layout-versus-schematic (LVS) checks were developed such that a design is not passed to manufacture or any other stage until verifiably passed and the "green light" given. Many CMOS foundries will not fabricate a design or require extensive waivers before allowing the tape-out of an IC with known DRC errors.

5. Conclusions

The design of digital, analog, and sensing ASICs is an exciting but complex field. The references and texts provided are intended as an introductory reading list. Of course, once a student has an idea of a sub-specialty they wish to pursue, they can delve far more deeply. Alongside introducing the different areas within ASIC design, some key industry standards are discussed. Students are encouraged to engage with ISO-9001 as early as possible. This is simply prudent planning for a career within robustly managed engineering projects. Finally, a design flow called the universal design methodology was briefly covered, which emphases the need for agreement as to requirements and scope, along with verification of the design throughout the lifecycle of a complex design.

Author details

Edward M.D. Fisher

Address all correspondence to: emd.fisher@gmail.com

Coda Octopus Products Ltd (Sonar Systems), Edinburgh, United Kingdom

References

[1] Daly D, Fujino L, Smith K. Through the looking glass—The 2017 edition: Trends in solid-state circuits from ISSCC. IEEE Solid-State Circuits Magazine. 2017;9(1. Winter):12-22

[2] Daly D, Fujino L, Smith K. Through the looking glass—The 2018 edition: Trends in solid-state circuits from the 65th ISSCC. IEEE Solid-State Circuits Magazine. 2018;10 (1. Winter):30-46

[3] Harpe P. Successive approximation analog-to-digital converters: Improving power efficiency and conversion speed. IEEE Solid-State Circuits Magazine. 2016;8(4. Fall):64-73

[4] Faggin F. The making of the first microprocessor. IEEE Solid-State Circuits Magazine. 2009;1(1. Winter):8-21

[5] Klein D. The history of semiconductor memory: From magnetic tape to NAND flash memory. IEEE Solid-State Circuits Magazine. 2016;8(2. Spring):16-22

[6] Theuwissen A. Better pictures through physics. IEEE Solid-State Circuits Magazine. 2010;2(2. Spring):2-28

[7] Huber J, Successful ASIC. Design the First Time Through. New York, United States: Springer. 1991. ISBN-13: 978-1468478877

[8] Sze S. Semiconductor Devices: Physics and Technology. 2nd ed. New Jersey, United States: John Wiley & Sons; 2001. ISBN-10: 0-471-33372-7

[9] Allen P, Holberg D. CMOS Analog Circuit Design. 3rd ed. Oxford, United Kingdom: Oxford University Press; 2012. ISBN-13: 978-0199937424

[10] Weste N, Eshraghian K. Principles of CMOS VLSI Design: A Systems Perspective. 2nd ed. London, United Kingdom: Pearson; 1993. ISBN-13: 978-0201533767

[11] Toumazou C, Lidgey F, Haigh D, editors. Analogue IC Design: The Current-Mode Approach. London, United Kingdom: Institution of Engineering and Technology (IET); 1990. ISBN-13: 978-0863412158

[12] Yuan F. CMOS Current-Mode Circuits for Data Communications. New York, United States: Springer; 2007. ISBN-13: 978-1441939999

[13] Fisher E. Principles and early historical development of silicon avalanche and Geiger-mode photodiodes. In: Britun N, Nikiforov A, editors. Photon Counting. 1st ed. Rijeka: InTechOpen; 2018. ISBN-13: 978-9535157755

[14] Durini D, editor. High Performance Silicon Imaging: Fundamentals and Applications of CMOS and CCD Sensors. 1st ed. Sawston, United Kingdom: Woodhead; 2014. ISBN-13: 978-0857095985

[15] Kaeslin H. Digital Integrated Circuit Design: From VLSI Architectures to CMOS Fabrication. Cambridge, United Kingdom: Cambridge University Press; 2008. ISBN-13: 978-0521882675

[16] Wakerly J. Digital Design: Principles and Practices. 5th ed. London, United Kingdom: Pearson; 2018. ISBN-13: 978-0134460093

[17] Weste N, Harris D. CMOS VLSI Design: A Circuits and Systems Perspective. London, United Kingdom: Pearson. ISBN-13: 978-0321269775; 2004

[18] Lynn P, Fuerst W. Introductory Digital Singnal Processing with Computer Applications (Revised Edition). Chichester, United Kingdom: John Wiley & Sons; 1998. ISBN-13: 978-0471976318

[19] Hauck S, DeHon A, editors. Reconfigurable Computing: The Theory and Practice of FPGA-Based Computation (Systems on Silicon). 1st ed. Burlington, United States: Morgan Kaufmann Publishing; 2010. ISBN-13: 978-9380931869

[20] Mayer-Baese U. Digital Signal Processing with Field Programmable Gate Arrays. 4th ed. New York, United States: Springer; 2014. ISBN-13: 978-3642453083

[21] Trimberger S. Three ages of FPGAs: A retrospective on the first thirty years of FPGA technology. IEEE Solid-State Circuits Magazine. 2018;**10**(2. Spring):16-29

[22] Hennessy J, Patterson D. Computer Architecture: A Quantitative Approach. Burlington, United States: Morgan Kaufmann Publishing; 2017. ISBN-13: 978-0128119051

[23] Ranachandran S, Digital VLSI. System Design: A Design Manual for Implementation of Projects and ASICs Using Verilog. New York, United States: Springer; 2007. ISBN-13: 978-1402058288

[24] Rajan S. Essential VHDL: RTL synthesis done right. F. E. Chicago, United States: Compton Co; 1998. ISBN-13: 978-0966959000.

[25] Thomas D. Logic Design and Verification Using SystemVerilog (Revised). CreateSpace Independent Publishing. ISBN-13: 978-1523364022; 2016

[26] Salemi R. The UVM Primer: A Step-by-Step Introduction to the Universal Verification Methodology. 1st ed. Boston, United States: Boston Light Press. ISBN-13: 978-0974164939; 2013

[27] International Organization for Standards. ISO-9001: Quality Management Systems: Requirements [Internet]. 2015. Available from: https://www.iso.org/iso-9001-quality-management.html [Accessed: 30-12-2018]

[28] International Organization for Standards. ISO26262: Road Vehicles—Functional Safety [Internet]. 2018. Available from: https://www.iso.org/standard/68383.html [Accessed: 02-01-19]

[29] Radio Technical Commission for Aeronautics (RTCA). DO-254: Design Assurance Guidance for Airborne Electronic Hardware [Internet]. 2000. Available from: https://www.rtca.org/content/about-us-overview [Accessed: 30-12-18]

[30] Zeidman B. Chapter 4: Universal design methodology for programmable devices (UDM-PD). In: Designing with FPGAs and CPLDs. 1st ed. Florida, United States: CRC Press; 2017. ISBN-13: 978-1138436442

The Semiconductor Business

ASIC Commercialization Analysis: Technology Portfolios and the Innovative Performance of ASIC Firms during Technology Evolution

Erik den Hartigh, Claire C.M. Stolwijk,
J. Roland Ortt and Wim P.M. Vanhaverbeke

Additional information is available at the end of the chapter

http://dx.doi.org/10.5772/intechopen.79647

Abstract

We examine the relationship between application-specific integrated circuit (ASIC) firms' technology portfolios and their innovative performance. This relationship is complex, and we hypothesize that it changes according to the stage of ASIC technology evolution. We test our hypotheses using a longitudinal dataset of 67 firms from the ASIC industry over the period 1986–2003. We find that ASIC technology evolution negatively moderates the effects of the size and diversity of the internal technology portfolio on ASIC firms' innovative performance. This implies that, in earlier phases of ASIC technology evolution, successful ASIC firms developed large and diverse portfolios to cope with technological uncertainty. During later phases of ASIC technology evolution, they tend to have relatively smaller and less diverse portfolios, and they focus on unique, protectable, and exploitable advantages.

Keywords: ASIC industry, technology portfolio, technology diversity, innovation strategy, technology evolution, innovative performance

1. Introduction

We examine the relationship between the size and diversity of ASIC firms' technology portfolios and their innovative performance as ASIC technology evolves.

For many technology-based firms, in-house developed technology is crucial for the creation of innovations and for capturing innovation returns [1–3]. In-house technology development enables firms to increase the complexity of their innovations, so it becomes difficult for competitors to imitate them [4]. It also enables firms to maintain secrecy and, in that way, to establish a lead time [5]. Especially, technology that is classified as "distinctive competencies" [6] and that forms the core of the firm's technological capabilities will mostly be developed in-house because of these reasons. In-house technology development also creates absorptive capacity, i.e., the knowledge that enables firms to better understand, source, and use external technology [5, 7, 8].

In this chapter, we focus on the effects of the size and diversity of the in-house ASIC technology portfolio on ASIC firms' innovative performance. The size of the portfolio reflects the firms' total efforts to develop ASIC technology in-house. The diversity of the portfolio reflects how firms' development efforts are spread over various ASIC sub-technologies.

The relationship between the size and diversity of firms' internal technology portfolios and their innovative performance is complex, and the results of the previous research have been conflicting. We contribute to this research by investigating the moderating effect of technology evolution. As a moderator, we use Abernathy and Utterback's concept of technology evolution of an industry [9]. They distinguish three evolutionary phases: the fluid phase, the transitional phase, and the specific phase. Currently, ASIC technology is in the specific phase, according to patent counts, and this is also indicated by the industry's technology trends. The exact evolutionary phase may differ per ASIC sub-technology: gate array technology is at the end of its evolution, standard cell technology is in the late specific phase, and PLD technology is also in the specific phase. A competing technology such as FPGA is earlier in the specific phase. Emerging technologies, whether they are labeled as ASIC or as competing with ASIC, are in the fluid phase.

In the early, fluid phase of technology evolution, technological uncertainty is high, and firms need to keep various development options open to cope with that uncertainty. They need to develop large and diverse technology portfolios that are useful for various technology development scenarios. In the later, specific phase of technology evolution, after a "dominant design" has been established [10], technological uncertainty is much lower. This means that there is less need for large and diverse technology portfolios. This enables firms to focus on those technologies they can best exploit.

Managers of technology-based firms need to know whether and when during the evolution of a technology, investments in large or diverse technology portfolios contribute to their firms' innovative performance. Building and maintaining such technology portfolios require large and risky resource investments, and it is therefore important to ensure the returns to these investments.

To study the effects of portfolio size and diversity on innovative performance during technology evolution, we developed a longitudinal dataset of 67 firms from the ASIC industry over the period 1986–2003. Our results support the moderating effect of technology evolution on the relationship between technology portfolio size and diversity and firms' innovative performance. Our findings contribute to a better understanding of the complexity of these relationships.

The practical implication is that ASIC firms need to adjust their technology sourcing strategy according to the phase of ASIC technology evolution. In earlier stages of technology evolution, investing in a relatively large and diverse technology portfolio seems to be a better approach to improve innovative performance. In later stages, focusing on a relatively smaller and specialized technology portfolio seems to improve innovative performance. ASIC firms that focus on multiple technology areas need to balance their technology portfolios across the areas: focusing for the late-evolution technologies and investing and diversifying for early-evolution technologies.

2. ASIC industry and technology

The ASIC industry is a part of the semiconductor industry that can be characterized as an independent market for design modules [11] that has been a driving force behind major technological breakthroughs in the semiconductor industry [12]. The history of this industry is well known: the inventions of the point contact transistor (by John Bardeen and Walter Brattain in 1947) and the junction transistor (by William Shockley in 1948) in the Bell Labs provided, together with the diffusion-oxide masking photo process (1954), the integrated circuit (1958), and planar technology (1959), and the foundations for the development of the global semiconductor industry [11, 13]. By 1961, it had developed into a worldwide billion-dollar industry [13]. Although the development of ASIC technology began at the end of the 1960s [14], it became popular in the 1980s [15]. In the 1980s it became possible to combine standard integrated circuits (ICs) into custom ICs that were tailored to particular systems or applications or Application-Specific Integrated Circuits (ASICs) [16].

The successful development of ASICs requires the knowledge and competencies of different types of firms [17]. As a result, it is an industry characterized by newcomers, strategic alliances, and mergers and acquisitions [17]. The dynamic patterns and the need for different knowledge and competencies make it an attractive industry for our type of research. The major firms currently active in the ASIC industry are Texas Instruments, Infineon Technologies, STMicroelectronics, Renesas Electronics, Analog Devices (which acquired Linear Technology in 2017), Maxim Integrated Products, NXP Semiconductors, ON Semiconductor, Qualcomm, and Intel [18]. All these major players are active in multiple segments of the semiconductor industry, ASIC being one of them.

Analogous to most semiconductor firms, ASIC firms initially worked according to the integrated device manufacturer (IDM) business model. They vertically integrated every aspect of chip production, from design to manufacturing, packaging, and testing [19]. In 1984, Xilinx was the first firm adopting a "fabless" business model, focusing on the design of ASICs, and outsourcing its manufacturing to other "IDM" firms [19]. Soon after, in 1987, TSMC adopted a pure foundry business model, focusing on the manufacturing [20]. During the 1980s, most firms still used the IDM model. By now, most semiconductor firms use the fabless model, although a few major ones, such as Intel and STMicroelectronics, still use the IDM model, still accounting for about 55% of the market [21].

The industry reports generally define three ASIC subsegments:

- Full-custom design: a circuit that is customized on all mask layers and is sold to one customer.

- Semi-custom design: a circuit that has one or more customized mask layers, but does not have all mask layers customized and is sold to one customer. This segment includes gate array and standard cell technologies [22], although standard cell is sometimes placed separately in between full-custom and semi-custom designs.

- Programmable logic devices (PLD): a circuit with fuse that may be programmed (customized) and, in some cases, reprogrammed by the user. This segment includes CPLD, SPLD, and PAL technologies. FPGA technologies have in the past been regarded as a special kind of PLD but are now generally considered as a technology that competes with ASIC (see, e.g., [23]).

The three categories contain different devices with the same system functionalities that can be programmed at different moments in the development process; by the vendor (for standard cell, gate array), by the designer, prior to assembly (for full custom); or by the user (for PLD). Programmable logic devices offer the cheapest solution for low volumes. If volumes are higher and exceed a few thousand units, gate arrays offer the best solution. Full-custom devices are the best choice when production volumes exceed hundreds of thousands.

For the total ASIC industry, we can define the various phases of technology evolution based on our data. Based on the technological developments of all firms together, starting with the first patents developed in the ASIC industry, we can put the *fluid phase* before 1991, the *transitional phase* between 1992 and 1997, and the *specific phase* after 1998. This is an industry-level metric, which is in line with the framework of Abernathy and Utterback [9] and Utterback [24]. **Figure 1** shows the evolution of the number of firms in the industry. Although this data runs only until 2003, we can clearly see a reduction in the number of firms, indicating a mature market.

Looking at the ASIC subsegments, we start with gate array technology, which existed in some form since the mid-1960s but did not capture a sizable share until around 1983 [15]. By the 2010s this technology was hardly applied anymore [25]. In the mid-1980s, the standard cell was implemented. The first type of programmable logic device was invented in the 1970s [26], but the technology became more popular in the 1980s, and the PLD submarket became one of the fastest-growing sectors in the semiconductor industry. **Figures 2** and **3** show the units

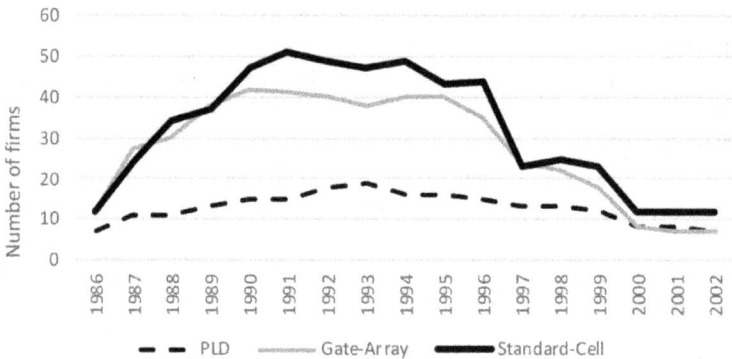

Figure 1. Number of ASIC firms.

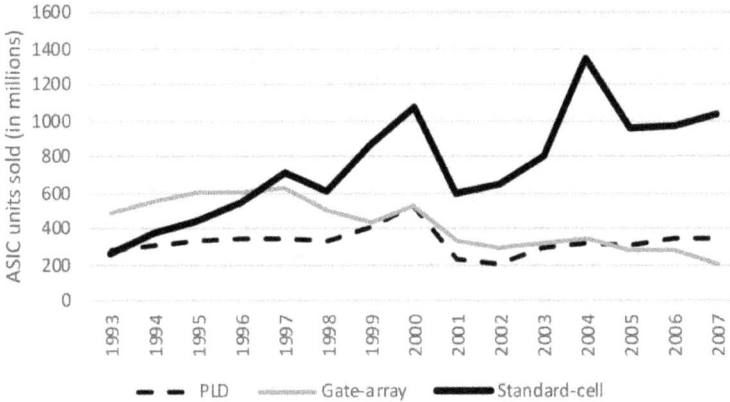

Figure 2. ASIC units sold (in millions).

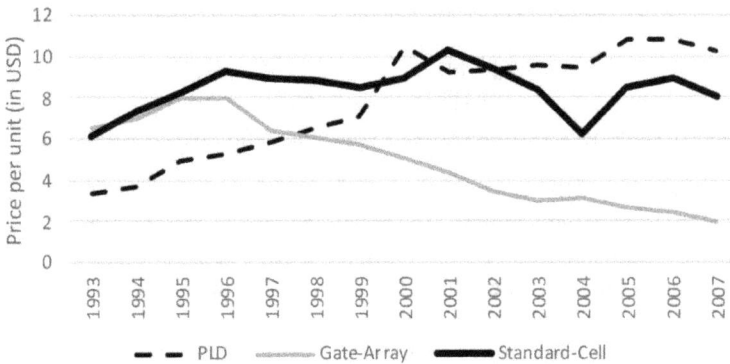

Figure 3. Price per unit (in USD).

sold and the prices per subsegment. These figures clearly indicate that the gate array technology evolution preceded the standard cell and PLD technologies in time.

We identify four major technology trends in the ASIC industry (see also [27, 28]). The first trend is the continuation of Moore's law through increasingly smaller DRAM pitch scales, increasing numbers of mask layers, and multi-patterning in lithography [29]. This trend is commonly referred to as "more Moore" [27, 28]. It entails strongly increasing cost of development of ASICs. As a consequence, minimum efficient design scale (numbers of ASICs sold per design) increases, and only few large design firms (fabless or IDM) are able to continue profitable operations, an indication of a mature technological field [9]. This development also fits the trend of firms concentrating on core competencies by adopting a fabless business model.

The second trend is increased efficiency in manufacturing due to wafer-size increases. The share of 300 mm wafers is still increasing [30], and efforts were made to increase wafer sizes from 300 to 450 mm [29]), although the consortium of firms involved abandoned its efforts at the end of 2016 [31]. This trend is mainly production process innovation, which like "more Moore" requires large investments in manufacturing facilities [29]. This, too, is a logical development in a mature technological field, and it fits in the trend of firms focusing on core competencies by adopting a foundry business model. Here, too, relatively few firms will be able to profitably carry out such investments because the minimum efficient scale of manufacturing ASICs increases.

These first two technology trends fit with market demand trends of ASICs as commodities for high-volume applications, such as Internet of Things, which require a lot of ASICs but not necessarily require leading-edge technology. Other markets with such demand are virtual and augmented reality, automotive electronics, smartphones, personal computing, and wearable electronics [32].

The third trend is added functionality, resulting in increasingly complex products for advanced applications such as machine learning or blockchain mining. This trend is commonly referred to as "more than Moore" [27, 28]. Examples of products are "software-defined hardware" or combinations of ASICs and general-purpose chips on a single-circuit board, like "domain-specific system-on-chip" [33] or "system-in-package" designs [27, 28]. This trend, too, indicates a relatively mature technological field with a focus on specific, albeit high-end, applications. This trend is accompanied by a shift from technology push-based roadmapping, to a more interactive approach in which multiple stakeholders are involved in defining future developments [27].

As a note on trends one till three: a mature technological field does not mean that technology does not develop or improve anymore. The technology still develops, e.g., in speed, power consumption, cost reductions, and performance for specific applications, but it develops in a relatively predictable direction and with a relatively predictable speed. This does not mean that such technology development becomes easier for firms: while the purely technological risks and uncertainties may be lower than before, the resource investments and the business risks are considerable, as are the business opportunities. In business terms, the industry moved from exploring new technologies to exploiting existing technologies.

The fourth trend, perhaps more accurately a collection of trends, is the emergence of new technologies such as quantum computing or nano-carbon technology [28]. This trend deals with completely new technological fields, and it is not always immediately clear whether these are still related to "ASIC" or the emergence of a completely new industry. Such uncertainty is a characteristic of the fluid phase of technology evolution [9]. This could be referred to as "beyond Moore."

3. Theory

Technological knowledge is a resource that helps create innovation by enabling firms to add value to incoming factors of production [34]. Here, we look at the size and the diversity of firms' technology portfolios. We would normally expect a positive effect of the *size of the technology portfolio* on innovative performance, because:

1. Technological knowledge embedded in patents is often converted into innovative products that contribute to firm performance (see, e.g., [35]). Given a certain efficiency of this function, more input (patents) will result in higher performance.

2. Knowledge as a resource is indivisible and self-generating, which cause it to have strong static and dynamic economies of scale in its application [3]. Indivisibility [36] means that a certain critical mass of technological knowledge is needed before it can be productively applied. Therefore, more technological knowledge can be expected to create higher innovative performance after this critical mass is reached. Self-generating ability [37] means that new relevant knowledge may emerge from the technology development process as additional output besides the normal output of (new) goods and services. The accumulated knowledge then becomes a basis for subsequent technological developments [7].

3. A larger technology portfolio allows for more recombination of knowledge (e.g., [38]). The possible number of combinations of knowledge exponentially grows with the size of the portfolio.

However, the relationship between technology portfolio size and innovative performance is more complex than expected. Lin et al. [39] find nonsignificant effects of technology portfolio (technology stock) on firm performance metrics. Artz et al. [40] show that, while the direct effects of R&D input to patent output (invention) and of patent input to product announcement output (innovation) are positive as expected, the indirect effect of R&D input on product announcement output is unexpectedly U-shaped and the indirect effect of patent input on firm performance is even negative.

The choice between a diverse and a focused knowledge base is one of the fundamental choices in a firm's knowledge strategies [1]. We would normally expect a positive effect of *technology diversity* on innovative performance because:

1. A diverse technology portfolio may generate economies of scope, or "synergies," meaning that it is more efficient to develop (related) technologies together than independently [3, 41].

2. Combining various technologies may generate "causal ambiguity," meaning that competitors are unable to determine the source of a firm's competitive advantage and therefore may have difficulty imitating it [4].

3. If we see innovation as a process of "recombinant search" for new combinations, a more diverse portfolio may result in many more possible combinations [38, 42].

However, this relationship, too, is more complex than expected. More diversity leads to increased coordination cost, which may partly or wholly offset the benefits, dependent on the strength of the firm's "integrative capabilities" [3]. Distributed technological capabilities may limit the firm's focus to develop strong core capabilities [43, 44]. The recombinant search advantage of a diverse portfolio depends on the degree of interdependency between components and on the size of the search space. Fleming and Sorenson [38] show that when interdependency is too high or too low, and when the search space is too large, recombinant search will become progressively less efficient. Building on this literature, Leten et al. [44] and Huang and Chen [45] argue that the relationship between diversity and innovative performance is complex and nonlinear. Lin [46] finds a nonsignificant relationship and suggests that diversity may interact with other variables.

A possible explanation for these complex results is that there is another variable that moderates the relationships between the size and diversity of a technology portfolio and innovative performance. We propose that technology evolution of an industry [9] is such a variable and that we may (partially) explain the complexities by including it in our model. We use Utterback's model [24] for our definition of technology evolution. This is a refined and validated version of the original Abernathy and Utterback model [9]. It specifies three phases in technology evolution: the fluid phase, the transitional phase, and the specific phase.

During the *fluid phase* of technology evolution, technological uncertainty is high. Technology solutions are not readily available, and technology development investments are explorative and focused on product innovation [9]. In this phase, firms in high-tech industries require technological scientific knowledge, i.e., knowledge gained through fundamental scientific research [47]. As a result of the uncertainty, firms do not know exactly which technological knowledge, i.e., which patents or combinations of patents, will improve their innovative performance.

They need to keep many options open and need to explore multiple different technological trajectories. To gain innovative performance in the early phases of technology evolution, firms need a large and diverse technology portfolio. A diverse technological knowledge base allows a firm to adapt [2] to turbulent environments and to develop a higher number of technologies. It also reduces the danger of a lock-in into dead-end technologies [48], and it hedges against the risks of developing the wrong technology [49]. As a result, diversification is positively associated with innovative performance. As technological scientific knowledge solutions are not readily available internally or externally, they need to be developed, adding to the portfolio size. Often, because of indivisibilities, various sub-technologies have to be developed simultaneously to create feasible technology solutions. When firms plan to source knowledge externally, they first need to develop a stock of knowledge internally that will enable them to scan and absorb external knowledge [7, 50]. During technology development, an increase in the number of components results in an exponentially larger number of possible combinations [38]. Grant [2] argues that different types of specialized knowledge are complements rather than substitutes, meaning that they are most useful when combined, or that there are economies of scope. A diverse technological knowledge base is required to be creative [47] and to create cross-fertilization between technological areas, which increases innovative performance [48]. Incidentally, this kind of cross-fertilization resembles the layout of the Bell Labs Murray Hill building in which the transistor was invented by design enabling—almost forcing—close contacts between researchers from different technological disciplines [51].

In the *transitional phase* of technology evolution, technological uncertainty decreases. Firms have had the chance to learn and acquire knowledge in the previous phase. Technological solutions are available, and technology development investments become more exploitative. In this phase, technology requirements shift toward application-related knowledge [47] and toward knowledge of process rather than product innovations. More certainty means that it is no longer necessary to explore many technological trajectories. In this phase, it is necessary to have a technology portfolio that is close to the dominant design. Therefore, in the transitional phase, firms require a smaller technology portfolio to gain innovative performance. To appropriate returns on technological knowledge, this knowledge should be unique to the firm, focused on the uniqueness of the portfolio not on portfolio size. When a firm needs technological knowledge outside of its own area of competence, there is a good chance that such knowledge is available with other firms and can be externally sourced. This also reduces the need to maintain large portfolios, provided the firm built up sufficient absorptive capacity in the fluid phase. Limiting the numbers of technologies generates cost advantages and thereby increases performance. Focusing the technology portfolio on the dominant design enables the firm to generate innovations that the market accepts, thereby increasing innovative performance. Having a unique technology portfolio close to its core competencies allows the firm to appropriate returns from the technology, which also leads to higher innovative performance. Since technology-related uncertainty is lower in this phase, firms can specialize rather than diversify their technological knowledge base, focusing on a narrow technological area [39] related to the dominant design. This creates important financial savings, which may be invested to improve the technological core, and in turn enables firms to outperform their rivals and maintain their technological leadership [39]. As much of the required technological scientific knowledge is available in this phase, either inside or outside of the firm's boundaries, the necessity to develop the scope of this knowledge is much lower. In this phase, it is more important to find the right applications for the knowledge that has been developed. Instead of being flexible and keeping all options open, firms should focus on their key technologies and core competencies [52]. This means increased specialization, leading to efficiency gains in knowledge acquisition and storage [34]. This applies especially when knowledge is specific to products or dominant designs because it is less subject to economies of scope than nonspecific knowledge [34]. During the transition phase, the cost aspect becomes more important, and it is too expensive to maintain a broad technological diversification.

In the *specific phase* of technology evolution, technological uncertainty is low, and most relevant technological knowledge is readily available. In this phase, firms need a small core technology portfolio to gain innovative performance. The dominant design is firmly established, and it is clear which technological knowledge is relevant. Since the technologies and products commoditize, cost savings are important, and maintaining a smaller portfolio will increase performance. During this phase, market-related rather than technology-related knowledge is required, and a large technological knowledge base is no longer necessary. As firms in this phase focus on exploiting existing knowledge, the uniqueness and protection of knowledge are even more important than in the transition phase. As it is not necessary to develop new technological knowledge, the economies of scope of a diverse portfolio no longer apply. It therefore makes sense to limit the technology portfolio to save resources. Any necessary related technological knowledge that is not available internally could easily be externally sourced. Saved resources can be invested in understanding the market and exploiting the firm's core technologies better.

In summary, we reason that during the fluid phase, firms need to develop large and diverse technology portfolios to cope with uncertainty and to keep development options open. In later phases, after a dominant design appears and technological uncertainty is lower, firms will benefit more from smaller, specialized portfolios that can be more easily protected and exploited. Based on this reasoning, we formulate the following hypotheses:

H1: *Firms with a large technology portfolio in earlier phases of technology evolution will achieve higher innovative performance than firms with a large technology portfolio in later phases.*

H2: *Firms with a diverse technology portfolio in earlier phases of technology evolution will achieve higher innovative performance than firms with a diverse technology portfolio in later phases.*

4. Data and methods

We test our hypotheses in the ASIC industry because it is knowledge-intensive, technology-intensive, and dynamic [11]. This makes it possible to measure the impact of the size and diversity of the internal technology portfolio on the innovative performance of ASIC firms during the phases of ASIC technology evolution.

We constructed a panel dataset that includes data from 300 ASIC firms and selected all 67 ASIC firms with innovative performance from this dataset for the period 1986–2003, using the Integrated Circuit Engineering ASIC-Outlook industry reports (1986–1999) and the Integrated Circuit Engineering status reports (1980–1999) until 1999. For the period between 2000 and 2003, we used the IC Insights reports [53], Compustat, product guides, Gartner reports, and data of the World Semiconductor Trade Statistics Corporation. We selected these firms based on the available data from the Integrated Circuit Engineering Corporation and IC Insights Company.

To measure our dependent variable, innovative performance, we collected data on the number of ASIC patents during 1986–2003 at t = i (where i = 1986, 1987, ... 2003). **Figure 4** shows an example of a firm's innovative performance over time.

We measured the first independent variable, the size of a firm's technology portfolio, by collecting data on the number of successful ASIC patent applications measured at t = i minus 5 (where i = 1986, 1987, ... 2003). We added all submarket-related patents per segment (PLD + gate array + standard cell) in 5 years prior to the year of observation. Henderson and Cockburn [54] recommend this moving window of 5 years, arguing that prior technologies can be expected to contribute to the development of new technologies. **Figure 5** shows an example of a firm's technology portfolio size over time.

We measured the second independent variable, the diversity of a firm's technology portfolio, by collecting data on the types of ASIC patents (PLD, gate array, standard cell) and by adding up all three submarket-related patents that a firm received during 5 years prior to the year of observation. This diversity is based on three types of technologies (PLD, gate array, standard cell) that the portfolios may contain in each year, which means that the value of this variable varies between 0 and 3: 0 if a firm has zero technologies in a year, 1 if a firm has one ASIC technology in its portfolios, etc. **Figure 6** shows an example of a firm's technology portfolio diversity over time.

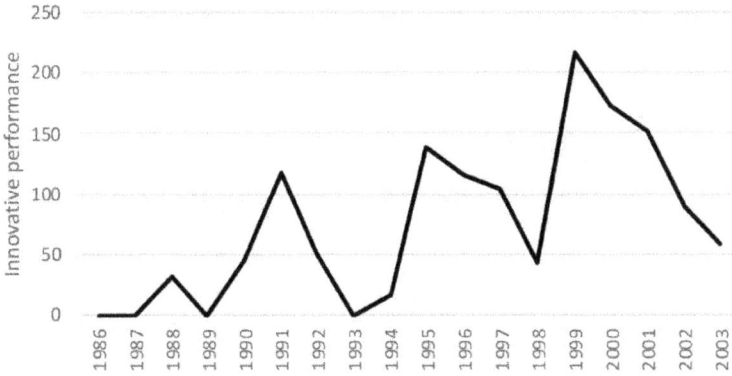

Figure 4. Example innovative performance.

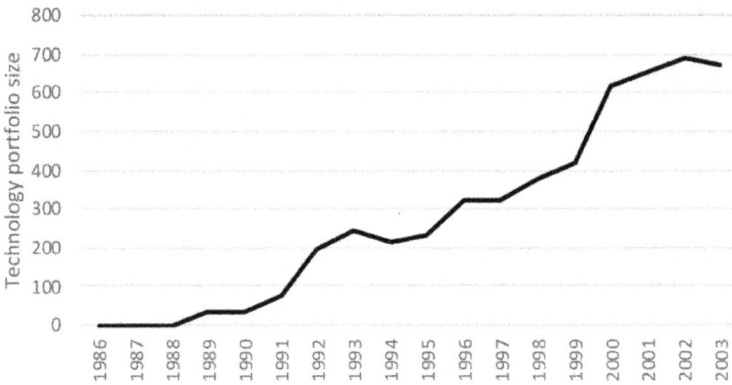

Figure 5. Example technology portfolio size.

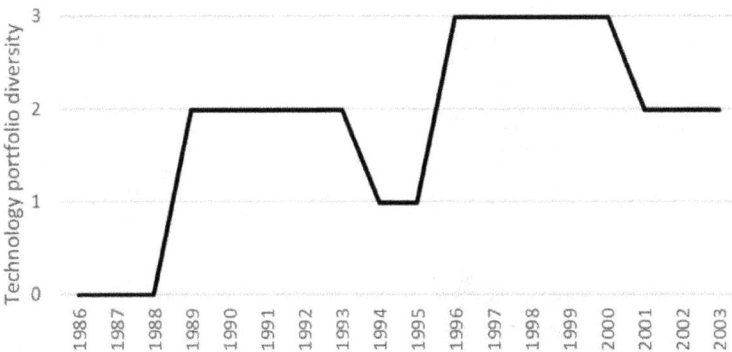

Figure 6. Example technology portfolio diversity.

We measured the moderating variable, technology evolution, based on its three phases: the fluid phase between 1986 and 1991, the transitional phase between 1992 and 1997, and the specific phase between 1998 and 2003. We use an industry-level metric, which is in line with the framework of Abernathy and Utterback [9] and Utterback [24]. It is based on the technological developments of all firms together, starting with the first patents developed in the ASIC industry. The metric is the same across the three subsegments, PLD, gate array, and standard cell.

We computed the interaction effect of technology portfolio size and technology evolution phase and the interaction effect of the technology portfolio diversity and technology evolution phase by multiplying the independent variables involved in the interaction. To enhance interpretability and eliminate nonessential multicollinearity, we standardized the independent variables in the interaction terms prior to computing those interaction terms [55]. We standardized the variables by first subtracting the overall mean from the value for each case, resulting in a mean of zero. We then divided the difference between the individual's score and the mean by the standard deviation, which results in a standard deviation of one.

We include five control variables. We measured the size of a firm's strategic alliance network as the number of cooperative relationships between firms: a firm's degree centrality [56]. We calculated this for each year using UCINET software. We use cooperation between firms over the last 5 years prior to the year of observation. We measured firm size as the natural log of the number of employees. Because larger firms are more dominant and have more financial

Variable	Time	Description	Mean	SD	Min.	Max.
Innovative performance	$t = i$	Number of ASIC patents per year	151	409	0	4764
Technology portfolio size	$t = i - 5$	Number of successful patent applications	590	1592	0	19,454
Technology portfolio diversity	$t = i - 5$	Zero, one, two, or three types of ASIC patents (PLD, gate array, standard cell) in the technology portfolio	1.35	1.24	0	3
Technology evolution	$t = i$	0 = fluid phase, 1 = transitional phase, 2 = specific phase	0.94	0.81	0	2
Alliance network size	$t = i - 5$	Normalized degree centrality	0.80	0.84	0	5.3
Firm size	$t = i$	Natural log of the number of employees	8.81	2.77	2.30	13.00
R&D/sales ratio	$t = i$	Percentage of sales invested in research and development	13%	15%	0%	300%
Asia region	$t = i$	Dummy variable denoting that the headquarters are located in Asia (default = America)	0.25	0.43	0	1
Europe region	$t = i$	Dummy variable denoting that the headquarters are located in Europe (default = America)	0.18	0.38	0	1

Table 1. Overview of the variables.

	Innovative performance	Technology portfolio size	Technology portfolio diversity	Technology evolution phase	Alliance network size	Firm size	R&D/ sales ratio	Asia region	Europe region
Innovative performance	1.00								
Technology portfolio size	0.81	1.00							
Technology portfolio diversity	0.32	0.37	1.00						
Technology evolution	0.20	0.31	0.32	1.00					
Alliance network size	0.22	0.26	0.46	0.28	1.00				
Firm size	-0.00	0.03	0.42	-0.02	0.27	1.00			
R&D/sales ratio	-0.01	-0.00	-0.11	0.03	-0.07	-0.43	1.00		
Asia region	-0.06	0.04	0.27	0.05	-0.19	0.35	-0.27	1.00	
Europe region	-0.14	-0.14	-0.09	-0.03	0.17	0.31	-0.05	-0.25	1.00

Table 2. Correlations.

means and resources to invest in R&D than smaller firms, they may have a higher innovation output compared to smaller firms due to economies of scale. We used a natural log because the number of employees is not normally distributed and the order of magnitude of the firm matters rather than its exact size. We used a firm's R&D expenses as a percentage of total sales, to check for the firm's propensity to invest in R&D. We also controlled for region indicating whether the firm's headquarters are located in America, in Asia, or in Europe.

Table 1 gives an overview of the variables. The firm-level data show a high average R&D intensity of 13% and a high average of 590 ASIC patents in the firms' portfolio, which indicate that the ASIC industry is knowledge-intensive and technology-intensive.

To test the hypotheses, we composed a longitudinal panel dataset. We conducted Hausman tests to decide whether to use fixed or random effect models. The panel analyses with the dependent count variable innovative performance are based on weighted patents. The mean and variance of the count distribution of these weighted patents are unequal, which means over-dispersion of the data, resulting in the need for a negative binomial regression [57].

Table 2 shows the correlation matrix. Based on the robustness of the test results (pair-wise exclusion of the variables with high correlations), no variables need to be excluded to avoid multicollinearity. Based on the results of the Hausman test, we selected fixed effects models for testing both hypotheses.

5. Results

To check our hypotheses, we test three models, the results of which are presented in **Table 3**. Model 1 is the baseline model that tests the direct effects of technology portfolio size and technology portfolio diversity on innovative performance. The model indicates positive and significant effects of portfolio size and portfolio diversity on a firm's innovative performance.

Model 2 tests how the technology evolution stage influences the relationship between technology portfolio *size* and innovative performance. It does so by including the interaction term of technology portfolio size and the phase of technology evolution. The estimates show that technology evolution negatively moderates the relationship between the size of the technology portfolio and the firm's innovative performance. It means that in later phases of the technology evolution, firms with smaller portfolios perform better. This supports Hypothesis 1.

Model 3 tests how the technology evolution stage influences the relationship between technology portfolio *diversity* and innovative performance. It does so by including the interaction term of technology portfolio diversity and the phase of technology evolution. The estimates show that technology evolution negatively moderates the relationship between the diversity of the technology portfolio and the firm's innovative performance. It means that in later phases of the technology evolution, firms with less diverse portfolios perform better. This supports Hypothesis 2.

All models indicate that larger firms and firms with larger networks have higher innovative performance, which is in line with findings of Gopalakrishnan and Bierly [58]. Larger firms

	Model 1 (main effects)	Model 2 (portfolio size and technology evolution)	Model 3 (portfolio diversity and technology evolution)
	Fixed effects♯♯	Fixed effects♯♯	Fixed effects♯♯
Technology portfolio size	0.000107*** (0.000)	0.000368*** (0.000)	0.000127*** (0.000)
Technology portfolio diversity	0.726*** (0.109)	0.586*** (0.110)	0.675*** (0.107)
Technology evolution phase	0.0403 (0.055)	−0.0056 (0.054)	0.274*** (0.071)
Technology portfolio size x technology evolution phase ♯		−0.357*** (0.036)	
Technology portfolio diversity x technology evolution phase ♯			−0.276*** (0.054)
Alliance network size	0.356*** (0.055)	0.350*** (0.054)	0.358*** (0.055)
Firm size	0.146*** (0.032)	0.164*** (0.032)	0.183*** (0.034)
R&D/sales ratio	−0.0162 (0.411)	0.065 (0.416)	0.0646 (0.408)
Asia region	0.772*** (0.147)	0.788*** (0.149)	0.763*** (0.147)
Europe region	−0.625*** (0.195)	−0.619*** (0.198)	−0.768*** (0.197)
Constant	−3.315*** (0.311)	−3.435*** (0.314)	−3.776*** (0.334)
Number of observations	807	807	807
Number of firms	59	59	59

Standard errors in parentheses. Significance levels: ***$p < 0.01$, **$p < 0.05$, *$p < 0.1$.

♯To calculate the interaction terms, we standardized the variables. For the main effects, the variables are not standardized.

♯♯The values of the Hausman test are for Model 1 Prob $> \chi^2 = 0.0042$, for Model 2 Prob $> \chi^2 = 0.0001$, and for Model 3 Prob $> \chi^2 = 0.0000$. Since the tests are significant ($p < 0.05$), the null hypothesis is rejected, and the fixed effects model is most appropriate.

Table 3. Results.

have larger knowledge bases, and firms with larger networks are able to attract more external knowledge, which can be complementary to internally developed technology. Given the positive and significant main effects, the effects of these two control variables are not surprising.

The models also show that R&D investments, measured as the R&D/sales ratio, have nonsignificant effects on innovative performance. While this may seem surprising, there are various possible explanations. First, in our data, R&D investment is measured for all the firm's technologies, not specifically for the ASIC technologies. Many of the firms in our dataset also develop non-ASIC technologies, so that only a part of their R&D investment is related to ASIC development. Second, the effects of R&D on performance have sometimes been found to be nonsignificant or curvilinear (e.g., [40]), and these effects are not captured in our model.

Related to this, R&D spending is regarded as an input to the development of a technology portfolio and may therefore be subject to the efficiency of the "invention production function" that is not captured in our model.

Finally, the models show that relatively more innovative firms were located in Asia and relatively fewer in Europe between 1986 and 2003. Explanations for this are that more ASIC-developing firms are based in Asia and fewer in Europe to begin with and that during this period some European firms exited the sector, whereas in Asia new players entered.

6. Discussion

The main effect of *portfolio size* is positive and significant in base Model 1 and remains so when we include the moderating effect of technology evolution in Model 2. Thus, firms with a larger portfolio show a better innovative performance, regardless of the phase of technology evolution. This is in line with earlier findings of Ernst [35], Fleming and Sorenson [38], and Granstrand [3].

We find that technology evolution negatively and significantly moderates the relationship between technology portfolio size and innovative performance. This is a possible explanation for the previous conflicting results of Lin et al. [39] and Artz et al. [32]. Our results indicate that it is more beneficial for a firm to have a relatively large portfolio in an earlier phase of technology evolution and to reduce the size of its portfolio in later phases. To put it differently, in the earlier phases, firms are more focused on production of knowledge from R&D, whereas in later phases, they are more focused on production of innovation from knowledge. Conducting cross-sectional research in an earlier phase would result in underestimating the production of innovations from R&D, while doing so in a later phase would result in overestimating the production of innovations from R&D.

If we return to the characteristics of technological knowledge as we mentioned before, namely, economies of scale [3], indivisibilities [36], and self-generative abilities [37], firms likely need to accumulate a certain critical mass of technological knowledge in earlier phases before such knowledge becomes productive and leads to innovative performance. Conversely, in later phases, when such critical mass has been reached, it should be easier to achieve innovative performance, and expanding the technology portfolio is unnecessary.

Our findings for the *portfolio diversity* are similar to those for portfolio size. Here, too, the main effect of portfolio diversity on innovative performance is positive for base Model 1 and for Model 3 that includes the moderating effect of technology evolution. This is in line with earlier findings of Granstrand [3] and Breschi et al. [59] that a diverse portfolio is associated with innovativeness.

We find that technology evolution negatively and significantly moderates the relationship between technology portfolio diversity and innovative performance. This finding complements existing explanations of the complexity of the relationship between technology portfolio diversity and innovative performance. Our research indicates that it is beneficial for a firm to have a relatively diverse portfolio in earlier phases of technology evolution and to reduce portfolio diversity in later phases.

It is widely recognized that this relationship is complex. Granstrand [3] argued that the coordination and integration costs of multidisciplinary R&D become higher with increased diversification. Research by Leten et al. [44] and Huang and Chen [45] confirms this argument. They found an inverted U-shaped effect of technological diversification on technological performance. While technological diversification enables combination and recombination, (too) high levels of diversification provide only marginal benefits due to high coordination and integration costs.

Our findings complement this explanation by arguing that more coordination efforts are needed in the earlier phases of technology evolution when technologies are unknown and that less coordination efforts are needed later when the relevant technologies are much better known. Therefore, we suggest that the inverted U-shape will have steeper slopes during earlier phases of technology evolution, when there are both high benefits from technology diversity and high costs of technology diversification. The inverted U-shape will have gentler slopes in later phases, when the benefits from technology diversity are less and the cost of technology diversification is lower.

Whether the firm can gain net benefits from the balance between technology diversity and coordination costs depends on the integrative capabilities of both technologists and managers [3]. If the firm possesses the capabilities to integrate diverse technologies, this is associated with causal ambiguity and sustainable competitive advantage [4].

7. Conclusions

The relationships between the size and diversity of firms' internal technology portfolios and their innovative performance are complex. We contribute to the literature by introducing technology evolution as a moderating variable of the relationship between internal technology sourcing and innovative performance. Our results support these moderating effects. The findings from our study contribute to explaining the complexity of the relationships between technology portfolio size and diversity and innovative performance by offering a possible explanation for conflicting empirical findings (technology portfolio size) and by offering an explanation that complements earlier findings (technology portfolio diversity).

Our findings suggest that during earlier phases of ASIC technology evolution, ASIC firms need broad technological portfolios and technological capabilities to keep their options open to adapt [2], to avoid lock-in [48], and to avoid investing in the wrong technology [49]. Such a broad portfolio requires strong integrative capabilities to profit from technology diversity. As such in earlier phases, causal ambiguity is created, making the firm's innovation difficult to imitate. During later phases of ASIC technology evolution, ASIC firms need to focus on their core technologies and their core capabilities [52], in which the causal ambiguity has been embedded. In these phases, the role of integrative capabilities would be less pronounced.

For managers in the ASIC industry, our results imply that they need to invest in a large and diverse technology portfolio in the early phase of technology evolution and need to maintain relatively smaller and less diverse technology portfolios later on, to optimize their firm's innovative performance. Having a large and diverse ASIC portfolio in early phases of technology evolution gives the firm the flexibility to keep all options open during uncertain periods, while a smaller and specialized portfolio contributes to a focus on the core competencies in more certain periods. In the fluid phase, ASIC firms need to explore the technology space by developing a large and diverse technological knowledge portfolio. In the transitional phase, they need to reduce the size and diversity of their technological knowledge base and focus on their own unique knowledge contribution within the dominant design, applying knowledge from their core technological base. In the specific phase, they need to concentrate on a small, focused, unique, protectable, and exploitable technological knowledge base.

ASIC technology is currently in the specific phase, and it therefore may make most sense for ASIC firms to focus on such a small, focused, unique, protectable, and exploitable technological knowledge base. As we argued in our discussion of the trends in the ASIC industry, they can do this by focusing on cost reduction and large-scale production of commodity products to earn back the ever-larger design and production investments or by focusing on providing added functionality solutions for specific high-end applications. Of course, while doing so, they need to separately manage their portfolios regarding emerging technologies. If they want to play an active role in such emerging technologies, they will need to develop large and diverse portfolios again to deal with the uncertainties that such technologies bring.

The research described in this chapter has several limitations, which can provide directions for future research. First, we tested the effects of the size and diversity of the technology portfolio separately. We recognize that the combination of both effects may have an impact on innovative performance as well. Lin et al. [39] suggest that firms with smaller knowledge stocks should concentrate on a specific technological field and that the size of the knowledge stock may moderate the relationship between diversification and performance. This implies that, for individual firms, there may be different roads to success: either building large and diversified technology portfolios (e.g., Intel or Texas Instruments) or developing small and focused technology portfolios (e.g., SK Hynix). Future research could investigate the implications of technology evolution for both these roads, e.g., by case study analyses. Second, we did not specifically include the interactions between internal and external sourcing through the innovation network. The past research indicates complementarities between internal and external technology sourcing (e.g., [8, 50]). This implies that firms could, for example, combine internally focused portfolios with external cooperation to ensure the necessary diversity. Further research could extend out model to include such effects. Finally, we did not include the effects of mergers, acquisitions, buyouts, and spin-offs as vehicles to manage and build technology portfolios. This, too, could be addressed by future research using case study analyses.

Author details

Erik den Hartigh[1*], Claire C.M. Stolwijk[2], J. Roland Ortt[3] and Wim P.M. Vanhaverbeke[4,5]

*Address all correspondence to: erik.denhartigh@ozyegin.edu.tr

1 Faculty of Business, Özyeğin University, Istanbul, Turkey

2 TNO Strategy and Policy, The Hague, The Netherlands

3 Faculty of Technology, Policy and Management, Delft University of Technology, Delft, The Netherlands

4 ESADE Business School, Barcelona, Spain

5 National University of Singapore, Singapore

References

[1] Bierly P, Chakrabarty A. Generic knowledge strategies in the U.S. pharmaceutical industry. Strategic Management Journal. 1996;**17**(Winter Special Issue):123-135

[2] Grant RM. Prospering in dynamically-competitive environments: Organizational capability as knowledge integration. Organization Science. 1996;**7**:375-387

[3] Granstrand O. Towards a theory of the technology-based firm. Research Policy. 1998; **27**:465-489

[4] Reed R, DeFillippi RJ. Causal ambiguity, barriers to imitation, and sustainable competitive advantage. Academy of Management Review. 1990;**15**:88-102

[5] Veugelers R, Cassiman B. Make and buy in innovation strategies: Evidence from Belgian manufacturing firms. Research Policy. 1999;**28**:63-80

[6] Granstrand O, Patel P, Pavitt K. Multi-technology corporations: Why they have "distributed" rather than "distinctive core" competencies. California Management Review. 1997;**39**:8-25

[7] Cohen WM, Levinthal DA. Absorptive capacity: A new perspective on learning and innovation. Administrative Science Quarterly. 1990;**35**:128-152

[8] Vanhaverbeke WPM, Belderbos R, Duysters G, Beerkens B. Technological performance and alliances over the industry life cycle: Evidence from the ASIC industry. Journal of Product Innovation Management. 2014;**32**:556-573

[9] Abernathy WJ, Utterback JM. Patterns of innovation in technology. Technology Review. 1978;**2**:40-47

[10] Anderson P, Tushman ML. Technological discontinuities and dominant designs: A cyclical model of technological change. Administrative Science Quarterly. 1990;**35**:604-633

[11] Dibiaggio L. Design complexity, vertical disintegration, and knowledge organization in the semiconductor industry. Industrial and Corporate Change. 2007;**16**:239-267

[12] Pieters M. Do alliance cliques matter? Explaining innovative performance and alliance network dynamics through alliance clique membership and technological capabilities [thesis]. Diepenbeek: Hasselt University; 2009

[13] Riordan M, Hoddeson L. Birth of an era. Scientific American. 1997;**8**(Special Issue The Solid-State Century):10-15

[14] Einspruch N, Hilbert J. Application Specific Integrated Circuit ASIC Technology. San Diego(CA): Academic Press; 1991

[15] ASIC Outlook Report. ASIC Outlook-An Application Specific IC Report and Directory, Integrated Circuit Engineering Collection; 1984

[16] Smith MJS. Application-Specific Integrated Circuits. Vol. 7. Reading(MA): Addison-Wesley; 1997

[17] Vanhaverbeke WPM, Duysters GM. A longitudinal analysis of the choice between technology-based strategic alliances and acquisitions in high-tech industries: The case of the ASIC Industry. In: Innovation in Technology Management-The Key to Global Leadership; Portland International Conference on Management and Technology (PICMET); 1997

[18] 360MarketUpdates. Global Application Specific Integrated Circuit (ASIC) Industry Production, Sales and Consumption Status and Prospects Professional Market Research Report 2017-2022 [Internet]. 2017. Available from: https://www.360marketupdates.com/global-application-specific-integrated-circuit-asic-industry-production-sales-and-consumption-status-and-prospects-professional-market-research-report–2017-2022-11026547 [Accessed: June 3, 2010]

[19] Sarma S, Sun SL. The genesis of fabless business model: Institutional entrepreneurs in an adaptive ecosystem. Asia Pacific Journal of Management. 2017;**3**:587-617

[20] TSMC Company Profile [Internet]. 2018. Available from: http://www.tsmc.com/english/aboutTSMC/company_profile.htm [Accessed: June 18, 2018]

[21] Hung H-C, Chiu Y-C, Wu M-C. Analysis of competition between IDM and fabless–foundry business models in the semiconductor industry. IEEE Transactions on Semiconductor Manufacturing. 2017;**30**:254-260

[22] Pal GP, Gupta M. Application-specific integrated circuits (ASICs). International Journal of Emerging Technologies and Engineering. 2014;**1**:40-44

[23] Erjavec T. Introducing the Xilinx Targeted Design Platform: Fulfilling the Programmable Imperative. Xilinx White Paper WP306; 2009

[24] Utterback JM. Mastering the Dynamics of Innovation: How Companies Can Seize Opportunities in the Face of Technological Change. Boston (MA): Harvard Business School Press; 1994

[25] IC Insights. McClean Report 2018 Content and Summaries [Internet]. 2018. Available from: http://www.icinsights.com/services/mcclean-report/report-contents [Accessed: June 3, 2018]

[26] Xilinx. Xilinx PLD Handbook [Internet]. 2006. Available from: http://www.xilinx.com/publications/products/cpld/logic_handbook.pdf [Accessed: March 28, 2012]

[27] Arden W, Brillouët M, Cogez P, Graef M, Huizing B, Mahnkopf R. "More-than-Moore" White Paper [Internet]. 2010. Available from: http://www.itrs2.net/uploads/4/9/7/7/49775221/irc-itrs-mtm-v2_3.pdf [Accessed: June 3, 2018]

[28] Bauer H, Veira J, Weig F. Moore's Law: Repeal or Renewal? McKinsey Global Institute; 2013

[29] ITRS. International Technology Roadmap for Semiconductors 2.0 2015 Edition Executive Report [Internet]. 2015. Available from: http://www.itrs2.net/itrs-reports.html [Accessed: June 18, 2018]

[30] IC Insights. IC Makers Maximize 300mm, 200mm Wafer Capacity. Research Bulletin [Internet]. October 12, 2017. Available from: http://www.icinsights.com/news/bulletins/ [Accessed: June 18, 2018]

[31] McGrath D. No Sign of 450 mm on the Horizon. EE Times [Internet]. January 13, 2017. Available from: https://www.eetimes.com/document.asp?doc_id=1331180 [Accessed: June 18, 2018]

[32] IC Insights. Market Drivers Report 2018 Introduction [Internet]. 2018. Available from: http://www.icinsights.com/services/ic-market-drivers/ [Accessed: June 3, 2018]

[33] Galeon D. DARPA: We Need a New Microchip Technology to Sustain Advances in AI [Internet]. 2017. Available from: https://futurism.com/darpa-we-need-a-new-microchip-technology-to-sustain-advances-in-ai/ [Accessed: June 3, 2018]

[34] Grant RM. Toward a knowledge-based theory of the firm. Strategic Management Journal. 1996;17:109-122

[35] Ernst H. Patent applications and subsequent changes of performance: Evidence from time-series cross-section analyses on the firm level. Research Policy. 2001;30:143-157

[36] Teece DJ. Economies of scope and the scope of the enterprise. Journal of Economic Behavior and Organization. 1980;1:223-247

[37] Glazer R. Marketing in an information-intensive environment: Strategic implications of knowledge as an asset. The Journal of Marketing. 1991;55:1-19

[38] Fleming L, Sorenson O. Technology as a complex adaptive system: Evidence from patent data. Research Policy. 2001;30:1019-1039

[39] Lin B-W, Chen C-J, Wu H-L. Patent portfolio diversity, technology strategy, and firm value. IEEE Transactions on Engineering Management. 2006;53:17-26

[40] Artz KW, Norman PM, Hatfield DE, Cardinal LB. A longitudinal study of the impact of R&D, patents, and product innovation on firm performance. Journal of Product Innovation Management. 2010;27:725-740

[41] Suzuki J, Kodama F. Technological diversity of persistent innovators in Japan. Research Policy. 2004;33:531-549

[42] Fleming L. Recombinant uncertainty in technological search. Management Science. 2001;47:117-132

[43] Prahalad CK, Hamel G. The core competence of the corporation. Harvard Business Review. 1990;**68**:71-91

[44] Leten B, Belderbos R, Van Looy B. Technological diversification, coherence and performance of firms. Journal of Product Innovation Management. 2007;**24**:567-579

[45] Huang Y-F, Chen CH-J. The impact of technological diversity and organizational slack on innovation. Technovation. 2010;**30**:420-428

[46] Lin B-W. Knowledge diversity as a moderator: Inter-firm relationships, R&D investment and absorptive capacity. Technology Analysis & Strategic Management. 2011;**23**:331-343

[47] Li Y, Vanhaverbeke WPM, Schoenmakers W. Exploration and exploitation in innovation: Reframing the interpretation. Creativity and Innovation Management. 2008;**17**:107-126

[48] Garcia-Vega M. Does technological diversification promote innovation? An empirical analysis for European firms. Research Policy. 2006;**35**:230-246

[49] Hatfield DE, Tegarden LF, Echols AE. Facing the uncertain environment from technological discontinuities: Hedging as a technology strategy. The Journal of High Technology Management Research. 2001;**12**:63-76

[50] Cassiman B, Veugelers R. In search of complementarity in innovation strategy: Internal R&D and external knowledge acquisition. Management Science. 2006;**52**:68-82

[51] Gertner J. The Idea Factory: Bell Labs and the Great Age of American Innovation. New York: Penguin Press; 2012

[52] Kessler EH, Bierly PE, Gopalakrishnan S. Internal vs. external learning in new product development: Effects on speed, costs and competitive advantage. R&D Management. 2000;**30**:213-224

[53] McClean B, Mata B, Yancey T. The McClean Report: A Complete Analysis and Forecast of the Integrated Circuit Industry; 1998-2008

[54] Henderson R, Cockburn I. Scale, scope and spillovers: The determinants of research productivity in drug discovery. RAND Journal of Economics. 1996;**27**:32-39

[55] Aiken LS, West SG. Multiple Regression: Testing and Interpreting Interactions. Thousand Oaks (CA): SAGE Publications; 1991

[56] Hanneman RA, Riddle M. Introduction to Social Network Methods. Riverside (CA): University of California; 2005

[57] Cameron AC, Trivedi PK. Econometric models based on count data: Comparisons and applications of some estimators and tests. Journal of Applied Econometrics. 1986;**1**:29-53

[58] Gopalakrishnan S, Bierly PE. The impact of firm size and age on knowledge strategies during product development: A study of the drug delivery industry. IEEE Transactions on Engineering Management. 2006;**53**:3-16

[59] Breschi S, Lissoni F, Malerba F. Knowledge-relatedness in firm technological diversification. Research Policy. 2003;**32**:69-87

Design Methods and Techniques

Case Study: First-Time Success ASIC Design Methodology Applied to a Multi-Processor System-on-Chip

Arya Wicaksana, Dareen Kusuma Halim,
Dicky Hartono, Felix Lokananta, Sze-Wei Lee,
Mow-Song Ng and Chong-Ming Tang

Additional information is available at the end of the chapter

http://dx.doi.org/10.5772/intechopen.79855

Abstract

Achieving first-time success is crucial in the ASIC design league considering the soaring cost, tight time-to-market window, and competitive business environment. One key factor in ensuring first-time success is a well-defined ASIC design methodology. Here we propose a novel ASIC design methodology that has been proven for the RUMPS401 (Rahman University Multi-Processor System 401) Multiprocessor System-on-Chip (MPSoC) project. The MPSoC project is initiated by Universiti Tunku Abdul Rahman (UTAR) VLSI design center. The proposed methodology includes the use of Universal Verification Methodology (UVM). The use of electronic design automation (EDA) software during each step of the design methodology is also presented. The first-time success RUMPS401 demonstrates the use of the proposed ASIC design methodology and the good of using one. Especially this project is carried on in educational environment that is even more limited in budget, resources and know-how, compared to the business and industrial counterparts. Here a novel ASIC design methodology that is tailored to first-time success MPSoC is presented.

Keywords: ASIC design, first-time success, design methodology, MPSoC, UVM

1. Introduction

The ever-changing and ever-challenging ASIC design environment consistently brings up new challenges into the ASIC design league. There are long existing challenges [1] such as soaring cost (design, verification, fabrication), high design complexity, tight time-to-market window and competitive business environment. The complexity of design and verification has grown tremendously in search for more computational power and capacity, which is also in accordance to the Moore's law. Nowadays the implementation of specific application such as face recognition in a System-on-Chip (SoC) has become a practice [2]. Another technology advancement is the development and use of Network-on-Chip (NoC) for communication inside the chip [3]. All of the advancement in technology today without doubt has enriched the ASIC products while at the same time introduced more and more challenges into the ASIC design and verification.

The long and many steps in the design and verification process of an ASIC may often lead to first silicon failure [4]. This is due to factors such as human errors, immature electronic design automation (EDA) software, lack of experience and discipline [5]. The first silicon failure is most often not acceptable in many business environments where a design re-spin is costly and may harm the continuity of the business. Hence, ensuring first-time success is fundamental to survive in the ASIC design league. The key success that is proposed here is the first-time success ASIC design methodology that describes the design and verification process in a well-defined manner starting from the user requirements (system specifications) until the fabrication (tape-out) of the ASIC. The proposed methodology also borrows the Universal Verification Methodology (UVM) [6] for the verification part of the design. The main objective of the proposed methodology is to serve as the guideline for achieving first-time success. The methodology has been used and proven for a Multiprocessor System-on-Chip (MPSoC) project initiated by the VLSI Design Center in Universiti Tunku Abdul Rahman (UTAR) as a demonstration of concept and viability.

The MPSoC project produces a first-time working silicon (first-time success) that is named RUMPS401. The RUMPS401 MPSoC is developed by using the proposed design methodology and with the use of relevant industrial standard EDA software tools (Synopsys, Mentor Graphics and Cadence). The first-time success RUMPS401 has proven that the proposed design methodology or the "UTAR first-time success ASIC design methodology" ensures first silicon success. Notably, the development of the RUMPS401 took place in educational environment where budget, resources and know-how are limited compared to the business and industrial counterparts. The novel UTAR first-time success ASIC design methodology is presented in the next section.

2. UTAR first-time success ASIC design methodology

The methodology is comprised of two major parts: front end and back end. Every development of a commercial electronic product is derived from the user requirements, which is translated into system specification. This system specification is the entry point in the proposed ASIC design methodology and the final design step of the methodology is tape-out for fabrication. At the front end part of the methodology, the end is the equivalence check process and the resulting netlist from the process is then passed for the back end step. The back end step starts with floorplanning using the netlist that has passed the equivalence check process earlier. The process is carried on through the final step that is the tape-out. Here is the proposed ASIC design methodology as shown in **Figure 1**.

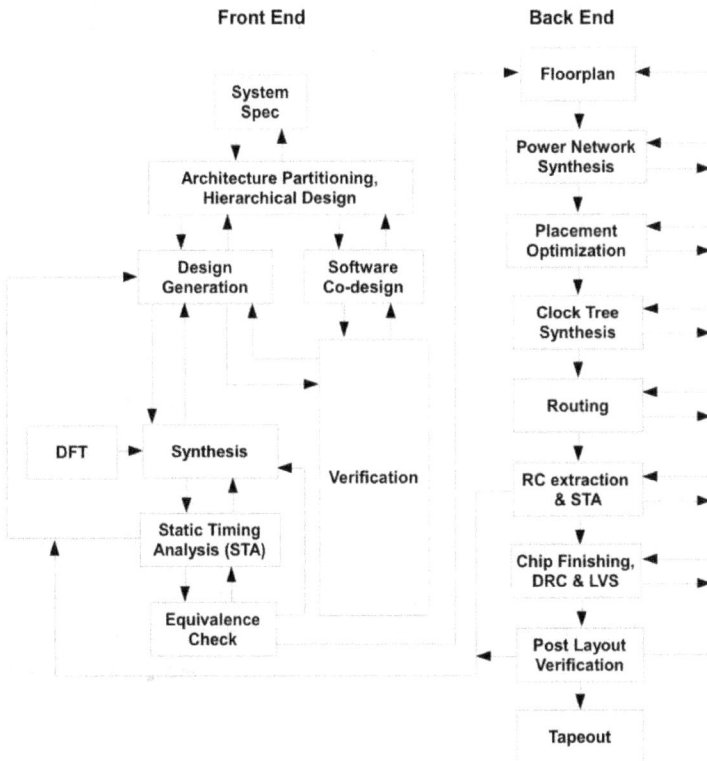

Figure 1. The proposed ASIC design methodology.

3. RUMPS401 MPSoC project

The objective of the RUMPS401 project is to be able to successfully implement the MPSoC to a ready-for-tape-out state. To fulfill this target, the "First-Time-Success" ASIC design methodology is strictly followed through the design flow. ASIC design methodology adoption affects the decisions in every design steps. This and the following sections elaborates the implementation of the methodology through the design process of the RUMPS401. In addition, the usages of the commercial ASIC design tools are discussed as well.

3.1. Design specification

The first design process of the RUMPS401 MPSoC is defining the system level functional specifications. RUMPS401 chip is targeted for microcontroller-based applications that perform relatively high data processing. To support this requirement, the chip is designed to contain multiple processor cores. RUMPS401 consists of four ARM Cortex-M0 processors. M0 processor implements Advanced High-Performance Bus (AHB) protocol to connect with the peripheral modules. Network-on-Chip (NoC) interconnect is selected as the inter-processors communication media. NoC architecture is preferred over the bus-based interconnect. It is more suitable for an MPSoC system due to it scalability property and capability to deliver better performance than it bus-based counterparts [7–10].

Several targeted applications—such as: coarse-grain Advanced Encryption Standard (AES) encryption and Software-Defined Radio (SDR)—are used to derive the functional specifications. For instance, in the coarse-grain AES encryption application, the targeted functional features are:

- The SoC retrieves a continuous stream of plaintext as the input.

- The plaintext are divided into data blocks and distributed to multiple processor cores through the inter-processors communication architecture.

- Multiple processor cores perform parallel coarse-grain encryption on the plaintext.

- The system accumulates the ciphertext as the result of the encryption process and arranges it in the correct order.

- The SoC transmits out the ciphertext as the output.

The design is partitioned into hardware and software in accordance to the functional specification. The hardware design of the RUMPS401 is partitioned into sub-systems level of design and furthermore into the modules level of design. Hierarchical design approach simplifies the design and verification process in the early stage of the design flow. Modularity design technique also enables the reusability of the modules in different part of the design. Some modules are designed specifically to enable and accelerate the targeted applications. For example, the AES accelerator and multiply-accumulate (MAC) module are targeted for the encryption and SDR applications respectively. The other modules such as the Flash and SRAM memories controllers support the basic functionality of the MPSoC. The defined specifications of all specific modules are summarized and written into specification document.

In the evaluation process of the design specification of the modules, several important aspects are factored into the analysis. It includes: performance, area and power consumption. Some key analyses are also discussed in the few next sub-sections.

3.2. Design partitioning

The RUMPS401 chip functional features are partitioned into smaller and more specific tasks. These tasks are then assigned to the four processor cores. For that reason, each of the processors has different sets of hardware modules. With this architecture implementation, RUMPS401 can be categorized as a heterogeneous MPSoC. One of the processor cores, called IO-control core, is designated to handle Inter-chip communication tasks. The core is equipped with communication modules such as parallel-port controller and has the largest number of IO pins. It is also utilized as the main coordinator core. The other cores are used for more specific functions. Two of the cores are named normal cores and equipped with AES accelerators. Likewise the DSP core contains MAC module.

The hardware implementation of parallel port controller, AES accelerator and MAC modules help the processor cores to perform repetitive tasks. In the application, these tasks especially take the most processing time. The processor cores are used to handle the less demanding tasks. Application software is designed to run the processor cores to execute these tasks. Scheduling and synchronization of tasks executions are also implemented in the software. These two aspects are increasingly important in the system that contains multiple processor cores.

The processor cores require AHB (Advanced High-performance Bus) to NoC (ahb2noc) module as the on-chip communication bridge. This ahb2noc bridge is specifically created to interface the M0 processors in the RUMPS401 that use AHB-lite protocol and the NoC inter-connect protocol. The data buffering processes in the ahb2noc are implemented in hardware to help speeding up the processor cores. Similar data buffering mechanism is applied to the parallel port module. These modules are used to transmit and receive multiple numbers of data. The modules are designed to register the received data in their buffer and generate inter-rupt signal to inform the processor cores about the availability of data. With this scheme, the processor cores are only needed to serve the Interrupt Service Routines (ISRs) once for some period of time. The mechanism helps the processor cores to save numerous processing times in compared with the mechanism without physical buffers.

RUMPS401 MPSoC does not contain any analog design block. If an analog design module is required such as Analog-to-Digital Converter (ADC), the development of the analog block is started together with the development of the digital hardware module and the application software. Behavioral model of the analog block is required to be prepared for the simulation purpose.

3.3. Network-on-Chip

NoC-based System-on-Chip is divided into multiple Processing Blocks (PBs). Each PB could contain one or a few components, and all of PBs are connected to each other via routers. A router connects to one or more neighbor routers depends on the topology of the network.

NoC used packet-switched method to send data from each PB to one another. All of the rout-ers have an exclusive address and routing algorithm. Data and header are wrapped as a flit. Flit header consist of destination address and type of the data. Each PB sends flits to destina-tion router through the network. The router's routing algorithm will determine the paths for the flit to reach its destination.

NoC has become the solution of choice for System-on-Chip traditional interconnects problem [11]. Each router is directly connected to neighbor router, as a result interconnect wire can be kept short. NoC interconnects are also able to handle multiple communication flows in the form of multiple flits and thus providing high communication bandwidth.

The design of NoC are divided in various aspects such as the topology, routing algorithm, flow control, and router microarchitecture. Topology determines physical layout, connections between nodes, channels in the network, and has an important role in routing strategy and application mapping [12]. Topology design starts with basic typical network topology for exam-ple, tree, ring, star, or mesh topology. Optimization and additional feature are implemented to

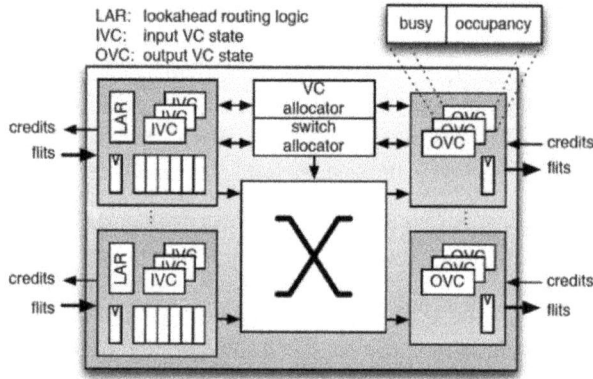

Figure 2. Router microarchitecture overview [15].

improve these basic topologies, as result new and more complex interconnects are formed. For example, the Quarc NoC architecture is an improved ring NoC topology [13]. Floor-planning and layout of the NoC architecture need to be considered as well [14].

Routing algorithm determines the path selected by a flit to reach its destination. A good routing algorithm is able to balance the traffic inside the network so that the performance and throughput of the network will be optimized. Flow control regulates the resource allocation process to a flit as they travel through the network. Congestion control, energy and power management and fault tolerance are communication design problem that need to be resolve as well [12].

Application modeling and optimization is another challenge in a NoC-based MPSoC. Every application has different communication traffic pattern. Different application mappings will significantly affect the network traffic. A good understanding of the traffic patterns and system requirements helps in determining the optimal network topology, and this has a huge impact on design costs, power, and performance [12].

Router performance, cost and efficiency depend on its microarchitecture [15]. Router micro architecture made up of input buffer, router state, routing logic, allocator and a crossbar. Input buffered routers have become the primary choice in current NoC research. Flits that cannot be forwarded are temporarily held in input buffers at the router until they can proceed to next hop. **Figure 2** shows a typical NoC router microarchitecture.

4. Front-end design

The RUMPS401 design is carried on in Register-Transfer Level (RTL). The Verilog Hardware Description Language (HDL) is used to implement the module blocks. The ARM Cortex-M0 processor cores RTL design is obtained through the ARM's University Program. The program allows the usages of the M0 processor in university research projects. The SRAM modules that are attached to each core are generated by using ARM's Artisan SRAM generator, while the Flash memories for program storages are proprietary to Silicon Storage Technology (SST).

The Widely used commercial ASIC design tools by Synopsys are utilized through the entire design process. Universal Verification Methodology (UVM)-based Verification Platform is constructed to verify the modules and the system level functionality. The RUMPS401 design is then synthesized into gate-level netlist. The design is targeted to use Silterra's 180 nm CMOS Technology. Equivalent checks and gate-level simulations are performed to the gate-level netlist against the original RTL design.

4.1. UVM-based verification platform

The verification platform is a structure consisting of testbenches. The testbenches are designed in a hierarchical structure. The most basic testbenches are created to verify individual RTL modules. As multiple RTL modules are combined together to form a more complex subsystem, the components of the module level testbenches can be reused to build the subsystem level testbench [11]. The RUMPS401 MPSoC system level testbench are the top most level of this hierarchical verification platform. The testbenches are built based on the UVM that used System Verilog Hardware Verification Language (HVL). System Verilog adopts the Object Oriented Programming (OOP) concept to provide more robust and reusable features to write testbenches as compared to Verilog based testbenches. The testbenches based on UVM concept are composed of reusable verification components that are called Universal Verification Components (UVCs).

4.1.1. UVC

An UVC is an encapsulated, ready-to-use, configurable verification environment for a specific interface protocol, design sub-module or a full system [6]. Each UVC is designed for a specific design or protocol. It generates stimulus, performs checking and collects coverage information. The RTL modules or systems that are verified are referred as Device-Under-Test (DUT). Some examples of UCVs that are designed to verify the RUMPS401 verification platform are: AHB-Lite, NoC and parallel port UVCs.

An UVC is comprised of sub-components that are called Agents. Inside a UVC, each agent handles a specific functionality but is still related to each other. These Agents have another level of sub-components: Driver, Monitor and Sequencer. Driver is an active component that drives the signals to stimulate the DUT. A series of stimulus to be executed by the driver is generated by Sequencer. These test stimuli are organized as items called sequences. Monitor acts passively by sampling the DUT signals.

4.1.2. Testbench architecture

Depending on the complexity of the DUT, a testbench architecture can contain one or more UVCs. It is not uncommon that some DUTs require other RTL module(s) to help the testbench to run the simulation. **Figure 3** shows the testbench architecture for the Parallel-port module. The RTL module of AHB-Lite bus is integrated to the testbench architecture together with the

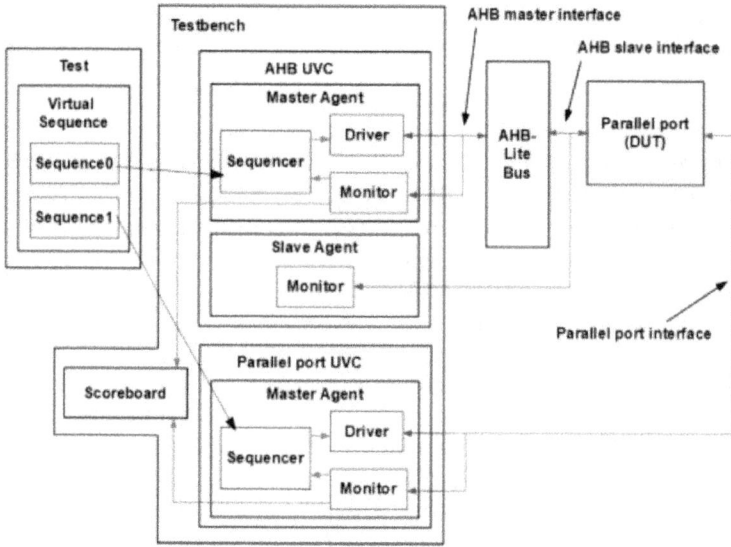

Figure 3. Parallel port module testbench architecture.

parallel port module. The AHB-Lite bus provides the necessary signals to access the AHB-Lite slave port of the parallel port module. An AHB-Lite Master Agent inside AHB-Lite UVC is connected to the AHB-Lite master port of the AHB-Lite bus module. The AHB-Lite Master Agent drives AHB-Lite transfers to access the registers inside the parallel port module. A passive AHB-Lite Slave Agent is attached to the connection between the AHB-Lite bus and the parallel port modules. This passive Agent only contains a Monitor and is only attached to collect the transfers. A parallel port UVC is connected to the parallel port interface of the parallel port module. The parallel port UVC trades data with the parallel port module using the parallel port data transfer protocol.

The test scenario to be run on the testbench architecture is defined in a component called Test. It selects the sequences to be run on all the sequencers inside the testbench to create a complete test scenario. When the test is running, scoreboard component is collecting all the necessary information. To obtain this information, scoreboard is connected to the monitors. The information are then used to perform comparison and check. Scoreboard reports failures if incorrect functionalities are performed by the DUT.

With the hierarchical approach, the module level testbenches are combined to form sub-system level testbenches. The sequences that generate some specific stimulus scenarios for the specific modules can also be reused. The same rules are applied when the sub-system RTLs are combined to form the MPSoCs [11].

4.1.3. Verification plan and coverages

Verification plan is constructed as the main guideline in the verification process. It contains multiple sets of tests for all possible test scenarios, especially all corner cases to stress the design. Some tests is designed as directed tests where the stimulus to drive the DUT are predefined. More advance tests generates constrained random stimulus. In this scenario, tests are randomized with the guide of constraints to make the stimulus to be more relevant.

There are two important goals in designing test scenarios. The first one is to test all possible functionalities of the module. The tests to cover these functionalities are derived from the RUMPS401 design specifications. 'Functional coverage' term is used to determine how much the functionalities of the design have been verified. Functional coverage reaches 100% when all of the functionalities are exercised. The second goal is to achieve a highest possible level of fault coverage. Ideally, good test scenarios are able to toggle all interconnect in the gate level netlist of the DUT and every failure must be observable at the output ports. However, due to the advancement of the geometry technology nodes that consistently increases the complexity of SoCs, fault simulation is no longer feasible. Code coverage concept is brought up as an alternative. Code coverage tracks how many line of code have been executed (line coverage), which paths through the code and expressions have been executed (path coverage), which single-bit variable have had the value 0 or 1 (toggle coverage) and which states and transition in a state machine have been visited (FSM coverage).

Verification coverage reviews are periodically performed through the design process. The verification coverage is applied as the parameter of the design progression. The periodic reviews are important for design evaluations and verification plan updates. Often new test scenarios are initiated as the result of analyzing the functional and code coverage reports. 100% verification coverage of the RUMPS401 chip is achieved towards the end cycle of the verification review as one of the signoff criteria.

4.2. Synthesis and design for testability

To ensure the synthesizability of the RUMPS401 RTL design, synthesizable coding guidelines are strictly followed. However, the synthesizability of an RTL module cannot be verified by simulation tools such as VCS. The Design Compiler (DC) tool from Synopsys is used to check design syntax and eliminate the non-synthesizable codes.

4.2.1. Design synthesis

The synthesis process is done to the Verilog RTL of RUMPS401 that are completed and fully verified. Design synthesis requires vendor specific synthesis libraries. The libraries contain information related to the available logic gates which are used as the target to translate and optimize the RTL design during the synthesis process. These libraries are prepared and loaded into DC tool. The RTL modules which are stored in multiple Verilog files are then read in and analyzed. DC reports errors when there are problems related to library or RTL synthesizability issues.

In order to guide the synthesis process, design constraints are applied to DC. The RUMPS401 design is targeted to operate with 16 MHz clock frequency. However for margin consideration, the design is constrained to 22.2 MHz. In addition 0.35 ns clock uncertainty constraint is applied to consider clock skew and jitter. For I/O-related timing constraints, time budgeting method is applied. For the chip input ports, DC requires information regarding the clock period and the output propagation delay of the signal from the device it interfaces with. For the outputs, other than the clock period, DC must also be provided with the input setup time specification of the device that is driven. With the time budgeting rule applied, 60% of the time allocation is dedicated to the external devices, and the remaining 40% is used to constraint internal IO controllers. The clock and reset signals are set as ideal nets therefore DC does not have to contend with clock skew issues at this stage. Clock and reset signals skew are particularly handled during the clock tree synthesis process in the physical design process.

In defining design constraints, cost consideration has to be balanced against performance. In the case where cost is the critical factor, the area constraint of the chip is set to some very small value or even zero. This constraint setup lets the DC to optimize the design as small as possible. In order to push the performance of the targeted technology, the clock period is set aggressively. Nevertheless sufficient timing margins must be included to consider physical variations, timing uncertainties and physical layout timing closure [11]. All these considerations are accounted to the limitation of the targeted 180 nm technology. Smaller process nodes allow tighter constraint such as higher system frequency and smaller clock margin to enable better performance. However smaller devices dimensions introduce worse leakage power that needs to be taken into consideration.

The final synthesized gate level netlist of the RUMPS401 design managed to meet the targeted constraints. There is a number of timing paths that violate setup requirement. However the slacks are in the range of 0.1–0.3 ns and are considered pretty small. These violations are left to be solved by the layout tools. The synthesized netlist requires 8.93 mm² total area. This value is the best case result that could be provided by DC due to the zero are constraint that is applied.

4.2.2. Design for testability

Physical defects can occur during the manufacturing process of chips. These physical defects could cause faults on any part of the chips. The design for testability (DFT) concept provides the capability to detect the occurrence of faults in all parts of the chips [11]. Most often for cost consideration, chips only provide limited number of I/O pins. DFT techniques are developed to facilitate the uses of limited I/O pins to apply test vectors and to verify the results. DFT is driven by the need to provide 100% controllability and observability.

The main DFT technique that is implemented in the RUMPS401 design is scan chain insertion. With this technique, all of the flip-flops in the design is replaced with scannable flip-flops that have an additional multiplexer at its input pins. In normal chip operation mode, flip-flops are interconnected through functional logic gates. During the scan mode, the output of a flip-flop is linked directly to the input of the next flip-flop. This scenario creates a long shift registers chain. For some cases where the chip has a large number of flip-flops, multiple scan chains are constructed to speed up the test process. RUMPS401 chip contains four scan chains. The start and the end points of these chains are brought to the I/O pins of the chip.

DC provides a feature to add scan chain into the design that is synthesized. With this option enabled, the scan chains are inserted to the gate level netlist of the RUMPS401 MPSoC. An additional file is also generated by DC that contains the detail information regarding the scan chains structure in RUMPS401. This file and the gate level netlist are used as the input to the Automated Test Pattern Generation (ATPG) tool. The tool analyzes the scan chain structure and generates all possible combination of test patterns. The Synopsys Tetramax ATPG tool is used for RUMPS401 test pattern generation.

Memory devices such as Flash and SRAM modules in RUMPS401 are designed to have very high density of memory cells. These memory devices are prone to suffer from physical defects. Studies have been done to classify faults that commonly appear in the memory devices. Some of the common memory related faults are: stuck-at faults, transition faults, coupling faults, bridging faults and state coupling faults. Test algorithms are developed to perform faults detection in memories modules.

A dedicated Memory Built-In-Self-Test (MBIST) circuit is designed to implement the memory test algorithms. The module is embedded into the SRAM controller modules. The MBIST module implements checkerboard algorithm, where it fills the entire array of memory cells with checkerboard pattern. The memory cells are alternately filled with one and zero values in vertical and horizontal direction. The MBIST module then reads back the values to detect any defect in the memory. The same test is repeated with the inversed pattern.

Similar to the MBIST, Flash wrapper interface is designed to provide direct access to the Flash memories in the RUMPS401. The Flash wrapper module implements multiplexer that is used to select one of the four Flash modules. The selected Flash module I/O pins are exposed by the Flash wrapper to the chip's I/O. Through the wrapper, test algorithms to erase, program and read the Flash memory cells are conducted. In addition, this Flash wrapper interface is also utilized to initialize the Flash modules after the manufacturing proses. The Flash initialization process must be done to setup the operating parameters of the Flash memories before it can be used to store program codes.

Other DFT methods are considered with the incorporate use of the M0 cores inside the RUMPS401 chip. One of the areas that are covered by this DFT method is the data buffers, such as in the NoC routers. Test software is written to be executed by the processor cores to pump data through the NoC buffers. The data is then read back and verified. The same method is also applied to perform the test to the peripheral modules, by looping back the output pins of the peripherals back to its inputs.

4.3. Equivalence check and gate-level simulation

With the rise of design complexity of SoCs, formality verification becomes critically important to ensure the correctness of every design step. The equivalence check is a common verification step in the design flow that utilizes EDA tools. This check is performed to verify the equivalence of designs that are implemented in different abstraction level. Therefore equivalence check is usually performed after the synthesis process is done and after the physical design implementation is finished.

In this RUMPS401 project, the equivalence check is done with the Formality tool from Synopsys. The gate-level netlist of the RUMPS401 from the synthesis process is used as the input. The original RTL design is also included as the reference design. The Formality tool implements mathematical techniques to perform logic circuits comparison of the RTL and gate-level designs. The tool then generates equivalence check reports base on the comparison results. A configuration script is required to guide the Formality tools to perform all the tasks.

Simulations are also conducted to the gate-level netlist of the RUMPS401 chip. These gate-level simulations are implemented as the double check to ensure the functionalities of the gate-level netlist are the same as the RTL design. The gate-level simulations utilize the same verification environment that is constructed to verify the RTL design. The RTL design is substituted with the gate-level netlist and then all of the test scenarios are applied and verified. The simulation results show all the generated logic gates of RUMPS401 design are function correctly.

4.4. Static Timing Analysis

DC tool provides embedded timing analysis feature to evaluate the logic path delays during synthesis process. However a more accurate Static Timing Analysis (STA) is needed to be performed to the gate-level netlist. Several factors such as variation of manufacturing process, supply voltage level and working temperature affects the timing characteristic of the cells and interconnects inside the chip. STA is conducted to verify that the design meets the timing requirements across the corner variation of Process-Voltage-Temperature (PVT) [11]. The RUMPS401 chip is targeted to operate with ±10% supply voltage variation (2.97–3.63 V) and in the environment with temperature ranged from −40 to 125°C.

STA is performed to the RUMPS401 design by using Prime Time (PT) tool from Synopsys. PT implements mathematical methods to calculate accurate timing delays. The gate-level netlist, vendor specific libraries and design constraints are read into PT. PT calculates the timing delays for all possible paths and uses it to perform the analysis. It generates the timing reports as the result. Critical timing paths are listed in these reports. One of the worst critical timing path is located at the output data bus of the Flash memory. The Flash memory module requires maximum of 33 ns time to generate the output data. With the design is constrained to use 22.2 MHz clock frequency, the output data bus of the Flash memory only lefts 12 ns time allocation. Refinements on the RTL design or synthesis process are done to reduce severe timing violations. In the Flash memory output data bus case, tight constraint is applied to the data path and the standard cells with the highest drive strength are used with the expense of chip area.

5. Back-end design

The back-end design process translates the logical RTL design into a layout, which defines how the gates and IP blocks are physically laid out in silicon wafer, as well as the connection between them. The end-result of this process is a layout file that foundry requires for the silicon fabrication. This section describes the back-end design steps done with Synopsys ICC tool and general view of the RUMPS401 physical layout design.

5.1. Initial design setup and floorplanning

Back-end design process starts by setting up the ICC with vendor-specific technology files and libraries. Technology files define the set of layers used in fabrication and design rules that must be followed. Libraries define the cells and IP blocks (logic gates) physical properties such as operating condition, gate delay, size, and pin locations. Additionally, the setup process also loads TLU+ files provided by vendor, which contain the parasitic RC model. Most importantly, the setup loads netlist and timing constraint files resulted from the design synthesis process.

The process continues by moving to the floorplanning step, in which the general layout or footprints of the chip is defined. This includes the total silicon area, core area, IO pads arrangement, power bus area, standard cells area, and IP blocks placement. The IP blocks placement is necessary as they usually have much larger size compared to the standard cells (gates) and are put on certain area to allow easier routing process. Normally, the IP blocks placement are done manually by the designer along with their blockages to prevent standard cells from being placed around the IP blocks, to reduce the congestion around the area.

Once the IP blocks are in place, standard cells are coarsely placed into the defined area. At the end floorplanning, the design will have all standard cells, IO pads, and IP blocks placed inside the defined silicon area. It is advisable to run multiple type of checks regarding congestion and core utilization. Core utilization is the ratio of area consumed by standard cells and IP blocks against total core area. Generally, it is kept at 40–50% to allow sufficient spaces for routing connections between the cells.

5.2. Power network synthesis (PNS)

Power is a crucial element for any chip as it is impossible to function without any electricity. The same goes for the cells inside the design. Hence, power network is laid out first to ensure that there is enough space for power lines, and that they utilize highest level of metal layer that is thicker than others for the best possible electrical characteristic. The goal of this step is to lay out power straps for the core area, from which the standard cells and IP blocks can tap into. In some design, due to operating voltage or current condition IP blocks may require separated power straps that tap directly into IO pads that supplies power.

Designer may choose to lay the straps manually or automatically with ICC's PNS feature. Either way, IR-drop analysis must be performed to ensure that the power straps is able to supply sufficient power to every cell in the design. It measures the voltage drop caused by strap resistance as a function of its length and width. The process of laying out straps and running IR-drop analysis is performed in cycles until the voltage drop level is within an acceptable range. In RUMPS401 design, the IR drop is kept below 50 mV across the chip. The chip uses 1.8 V CMOS technology for its internal cells and designed with ±10% tolerance for supply voltage, hence the maximum allowed IR drop is kept at lower value to accommodate the supply voltage variations.

The RUMPS401 power network are laid on the two highest layers of Silterra's six metal layers technology. Ground straps are laid on the sixth metal layer, while the 1.8 V straps lie on the fifth metal layer. The main power rings and are 100 and 30 µm wide, respectively. Standard cells power busses tap directly into the power straps through vias. There are 6–7 pairs of power straps laid vertically throughout the cell area to provide even power distribution and IR drop. Instead of tapping to the power ring or straps, the flash memories are powered individually via the power IO pads spread on every side of the chip. This is done in accordance to vendor's application note.

IR drop simulation were run separately for the 1.8 and 3.3 V supplies. Worst IR drop on the 1.8 V line is observed on the power straps going into the bottom-right flash, measure at 33.5 mV. An IR drop of 25.2 mV is measured on the 3.3 V power straps going into the top-right flash. Both are caused by longer distance from the IO pads, but are still below the 50 mV cap. As for the standard cells, the worst IR drop is measured at about 20 mV, located around the lower part of the chip. This is again due to the farther distance from the LDO located at the chip topmost. Verification of design constraints such as timing is performed after every step to ensure that there are no violation carried to the next design stage.

During the RUMPS401 CTS process, the clock optimization command was run in few optimization stages based on the verification result. After the initial CTS step, an optimization is performed for congestion and DFT, which result are checked with the global route congestion map. Re-optimization towards congestion were not performed as there is almost no congestion after the CTS. At this stage, the hold time constraint at 16 MHz clock path was violated by 0.02 ns and fixed by running the CTS optimization with the hold time prioritized, followed by the actual clock nets routing. Check on the congestion and timing were performed between every stage.

5.3. Placement optimization and clock tree synthesis (CTS)

At this point, the design physical layout has its standard cells, IP blocks, IO pads, and power straps in place. Placement of standard cells is optimized with ICC automation feature, in which the optimization parameters such as timing critical paths can be specified. The optimization groups together cells that belong to certain modules. It is intended to minimize the length of connection among cells, to reduce congestion, and to make the design routable.

Clock tree is a clock network that sources from a single IO pad and branches to every flip-flop in the design. Clock tree synthesis process attempts to generate the network with minimum clock skew, in which the clock signals arrived on every flip flops as evenly as possible. If required, buffers are placed to minimize clock delay on longer wire. This step is carried by running ICC automated CTS feature which works based on the DC timing constraint. The RUMPS401 clock tree is synthesized with 0.35 ns clock uncertainty derived from the design constraint defined during netlist synthesis. The CTS process is split into three steps. In the first step, the clock tree is synthesized without any optimization preferences. The second step optimizes CTS in terms of cell congestion and DFT. This step includes 1–2 rounds of optimizations to fix hold time violation. The actual routing of clock tree nets is performed in the last step, preceded by synthesis of reset signal using the CTS tool with the same constraints as the clock signal and target skew of 0.05 µs.

5.4. Routing

In this step, real wire connection among cells are made based on the logical connection defined in the netlist. Connections are formed via metal layers whose number depends on the foundry process. Prior to the ICC automated routing process, designer is required to specify certain parameters such as the routing priority, routing computational power. The automated routing process is done in three incremental steps. Leakage-power is the optimization target for the initial routing step. The second routing step optimizes further by attempting to optimize crosstalk reduction, while the final step cleans up any design-rule-checking (DRC) violations. DRC defines set of rules to be adhered by designers, such as minimum metal width, metal-to-metal spacing, antenna.

After every routing step, timing checks must be performed again to ensure that the timing condition is still met. Should it be violated, the routing may be carried again with timing optimization as priority. Generally, the design may not be clean of DRC violations after the first iteration of routing. The third routing step may be performed iteratively until the DRC is clean. Should there be violations that cannot be cleaned by the automated routing, they must be fixed manually by the designer.

At any point of the routing process, the physical design can be exported into physical netlist which is used for logical verification against the front-end design as well as timing analysis. Should there be a problem, changes are made on either front-end or back-end part. It may be big changes yielding to the repetition of whole back-end process, or small changes that can be solved by Engineering Change Order (ECO) process.

5.5. Engineering Change Order

This step is commonly performed when the back-end design is nearing its final step and changes in the design is minor. ICC provides ECO process where the changes can be accommodated by spare cells prepared by front-end designer. These cells are already defined in RTL level and takes form of common standard cells which are placed inside the core area but has no connection other than power. Should an ECO process occur, ICC will perform place optimization and minor re-routing to connect the design to these spare cells based on the ECO netlist.

5.6. Post-layout Static Timing Analysis, equivalent check and simulation

After the physical design of the RUMPS401 MPSoC is completed, Resistance and Capacitance (RC) characteristics can be extracted from the metal routes. These RC numbers provide more accurate timing delays across the signal paths. Another round of Static Timing Analysis is performed to consider these RC delays information. Adjustment is done to the physical design to eliminate any reported timing violations. In the first STA run, many hold time violations are reported, especially in the scan chain test mode. These are due to the direct path between the output pin of the flip-flops to the scan input of the next flip-flops. The parasitic RC delays affect the clock skew and the path delays. However, the violations are very minor and can be in the fractions of nanoseconds. Delay cells insertions are done to fix these violations. After all violations are fixed, the final timing reports are then used for design sign-off.

In this stage, equivalence check is required to ensure the whole back-end design processes do not alter the functionality of the chip. The post-layout netlist of RUMPS401 is extracted from the physical design. It is analyzed and compared against the gate-level netlist from DC as the reference design.

Post-layout simulations are performed as the final verification process of the RUMPS401 design. The post-layout simulations have very similar setup to the gate-level simulations. The extracted post-layout netlist is used as the DUT. RC characteristics of the design are also included to the simulation. Post-layout simulations verify that the design is still able to perform all the specified functionality with the additional parasitic delays added to the signals path. The final post-layout netlist of RUMPS401 chip with the parasitic timing delays manages to pass all the functional tests.

5.7. Chip finishing

To prepare the design for tape-out, there are few necessary finishing steps that must be applied to the physical layout. Filler cells are inserted into the layout to fill gaps among standard cells due to the manufacturing process requiring continuity of silicon within certain amount. IO pads fillers are inserted as well for the power ring continuation along the IO pads. To improve the routing wire quality, route tracks are widened and redundant via are created. After the finishing, various checks are run against the layout to ensure that it is free of any violation such as timing and DRC. The clean layout is then exported into a layout file (GDSII). Up to this point, the back-end design processes are performed in ICC tool environment.

The next step in chip finishing are performed Cadence's Virtuoso and Calibre DRC/LVS tools. They are required to produce a complete GDSII layout file which contains the detail of every silicon masking layer as well as their geometry, which the foundry will refer to for actual fabrication. Virtuoso is used to perform merging between the ICC-exported GDSII and the complete layout of every cell, IO pad, and IP block used in the design. In ICC, those instances are abstracted as block with IO pins to reduce the processing load required by ICC during place and route process.

The layout is then checked against DRC rules with Calibre tools. This DRC check is generally the same as ones performed in ICC, but Calibre has been widely used as industrial standards hence it is more common to find foundries requiring designer to do the final DRC check with Calibre ruleset. Layout-versus-schematic (LVS) check is run to compare netlists produced at the front-end and back-end for their logical equivalence. Once the layout passes the checks, it is ready to be sent over to foundry for fabrication.

5.8. General view of the RUMPS401 physical layout

Figure 4 illustrates the RUMPS401 chip and physical layout in general. SRAM and flash are two largest IP blocks in the design, and they are placed on each end of the core area. Below and above the pairs are arrays of power switches, used for controlling power going into the flash. The RUMPS401 allows independent core sleep, in which it will turn off the flash to

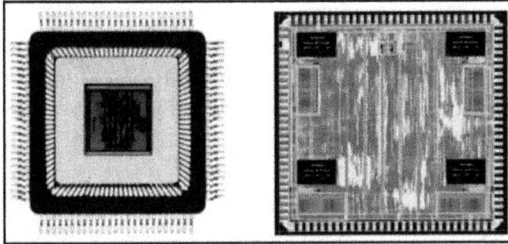

Figure 4. RUMPS401 chip (left) and RUMPS401 physical layout (right).

conserve power. Main power bus runs around the core area, supplying 1.8 V power source for standard cells. The 1.8 V is sourced from an internal 3.3 V-to-1.8 V low dropout voltage regulator (LDO) that receives 3.3 V power externally via IO pads. The 3.3 V power source is also used directly by the flash memory.

6. Result and application

Throughout the first-time success ASIC design methodology, check is performed on each design step to ensure that the design is free of error all the way to the fabricated silicon. When the physical chip is produced, proper testing procedure is performed first to verify that every gate in the silicon works, which includes but not limited to flash, RAM, and scan chain test. All gates are functioning verified by the scan-chain test. During normal operation, the RUMPS401 runs on the 16 MHz external clock source, consuming about 30 mA with all the cores running. The IO control core, normal cores, and DSP core individual current consumptions are at around 4, 7, and 4 mA, respectively. The NoC itself consumes about 10 mA and remains functional if at least one core is active. In a deep sleep mode all cores are put to sleep, the RUMPS401 switches to internal 32 kHz clock source and consumes only 13 μA. Ultimately, the goal of this thorough first-time success methodology is to have the chip working and performing its functionality on the first fabrication attempt. Hence, it is without doubt that the chip has been tested for real world application as a proof to the design methodology.

6.1. Software-defined radio

Software-defined radio (SDR) has been widely accepted as a solution for accommodating the availability of various yet ever evolving wireless standards. By shifting signal processing to programmable digital processors, a single radio transceiver can be reconfigured to operate on various standards without the need of hardware change. This is a desirable property for numerous applications, including Internet of Things (IoT). Industry players are competing on establishing wireless standards aimed for IoT, such as LoRa-WAN, SigFox, Bluetooth LE. During its lifespan, an IoT system may want to migrate to different wireless standard. As the number of nodes grow, replacing the transceivers would be problematic.

The RUMPS401 unique architecture offers an interesting platform for SDR implementation. Its low power and multi-processor architecture offers considerable processing capability if utilized properly whilst keeping power consumption to the minimum. Paired with a programmable radio frontend, the RUMPS401 can function as a complete IoT node, managing both SDR operation and transducers control. In [16], an SDR platform consisting of the RUMPS401 and programmable radio Lime LMS6002D was introduced. Implemented on the platform is an initial-stage transceiver software operating Turbo-coded, coherent binary phase-shift-keying (BPSK) modulation. Noting the absence of signal processing hardware from the RUMPS401 either in the ARM M0 processor or the peripherals, running coherent PSK modulation-demodulation and Turbo Code decoding is a challenging task for the MPSoC.

The Turbo Code software is implemented by dividing the algorithm based on its main tasks, i.e. the metrics calculation. Exploiting the heterogenous architecture, the tasks are split across the four cores based on each core's main functionality. The IO-control core is tasked with interfacing against the Lime LMS6002D and governs the sequence of data operation. The DSP core handles the computation of metrics that require extensive multiplications, while the two normal cores handle the rest of simpler but data-extensive metric computation. It was shown in [16] that such task structuring yields faster decoding compared to structure where each core performs more than one type of task due to the increased complexity on packet transfers among cores.

The coherent BPSK is configured to operate at a gross data rate of 2 kbps on 433 MHz carrier. Turbo encoding, pilot-bit stuffing and pulse shaping are applied to the data before transmission. The receiver performs pilot-aided timing and frequency synchronization on the incoming signal, then present the received data after passing it through the Turbo decoder. These functionalities are performed solely in the RUMPS401 via software, while the Lime LMS6002D only handles operations in radio frequency (RF) region.

Figure 5 illustrates the test setup with two SDR platform, one each for transmitter and receiver. Tests was carried in line-of-sight (LoS) configuration, in an empty room without any special conditioning. Excluding the pilot and Turbo code's parity bits, total of 10,000 pure data bits were generated and transmitted by the transmit side and checked on the receiver for its error rate. It is evident from **Figure 6** that the transceiver system exhibits the expected error rate behavior, as well as the Turbo Code that improves error rate by 20–40% on its third iteration.

This SDR platform has highlighted one of vast possibilities offered by the RUMPS401 as a low-power MPSoC on complex applications and serves as a proof of the first-time success

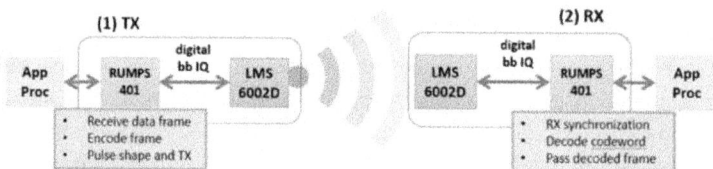

Figure 5. RUMPS401 SDR-based transceiver setup.

Turbo-coded System Test and Comparison (BER vs Transmit Range)

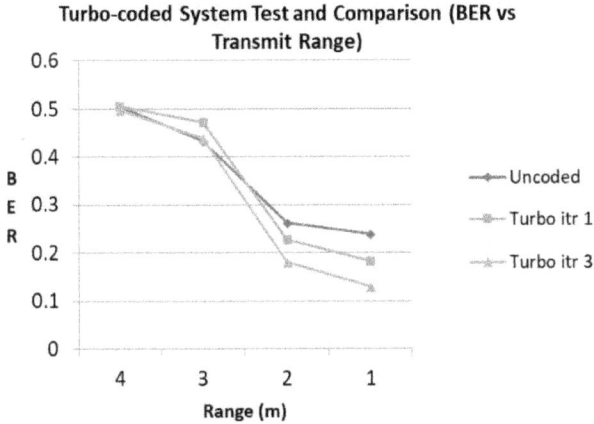

Figure 6. RUMPS401 SDR-based transceiver test result.

ASIC design methodology embraced by UTAR VLSI Research Center. Another example of RUMPS401 application is AES encryption as presented in [17]. The AES application has been successfully tested on the virtual prototyping platform and the RUMPS401 chip.

7. Conclusion

The first-time success RUMPS401 MPSoC has proven the proposed ASIC design methodology. The RUMPS401 is designed, implemented, and tested following the UTAR first-time success ASIC design methodology as presented throughout the book chapter. Design and verification processes are well defined and guided by the methodology which leads to first-time success. Corresponding EDA standard software tools are also mentioned including the use of UVM for functional verification.

Author details

Arya Wicaksana[1]*, Dareen Kusuma Halim[2], Dicky Hartono[3], Felix Lokananta[4], Sze-Wei Lee[4], Mow-Song Ng[3] and Chong-Ming Tang[3]

*Address all correspondence to: arya.wicaksana@umn.ac.id

1 Department of Informatics, Universitas Multimedia Nusantara, Indonesia

2 Department of Computer Engineering, Universitas Multimedia Nusantara, Indonesia

3 GLX Technologies, Malaysia

4 Department of Electronic Engineering, Universiti Tunku Abdul Rahman, Malaysia

References

[1] Kaeslin H. Digital Integrated Circuit Design. New York: Cambridge University Press; 2008. pp. 632-665

[2] ExtremeTech. What to Expect From Apple's Neural Engine in the A11 Bionic SoC [Internet]. 2017. Available from: https://www.extremetech.com/mobile/255780-apple-neural-engine-a11-bionic-soc [Accessed: 2018-06-03]

[3] Lusala AK, Manet P, Rousseau B, Legat JD. NoC implementation in FPGA using torus topology. In: 2007 International Conference on Field Programmable Logic and Applications. 2007

[4] Kharchenko VA. Problems of reliability of electronic components. Modern Electronic Materials. 2015;1:88-92. DOI: 10.1016/j.moem.2016.03.002

[5] Delta. SEM analysis reveals real cause of chip failure. Available from: https://asic.madebydelta.com/portfolio/sem-analysis-reveals-real-cause-of-chip-failure [Accessed: 2018-06-03]

[6] Accellera. Universal Verification Methodology (UVM) 1.1 User's Guide. California: Accellera; 2011

[7] Hartono D, Yap VV, Lee SW, Ng MS, Tang CM. A multicore system using NoC communication for parallel coarse-grain data processing. In: New Media Studies (ConMedia). 2013

[8] Dally WJ, Towles B. Route packets, not wires: On-chip interconnection networks. In: Proc Des. Autom. Conf. 2011. pp. 684-689

[9] Benini L, Micheli GD. Network on Chips: Technology and Tools. San Francisco: Morgan Kaufmann; 2006

[10] Wicaksana A, Tang CM. Virtual prototyping platform for multiprocessor system-on-chip hardware/software co-design and co-verification. In: Lee R, editor. Studies in Computational Intelligence. Vol. 719. Cham: Springer; 2011. pp. 93-108. DOI: 10.1007/978-3-319-60170-0_7

[11] Hartono D. The design and implementation of a scalable multi-processor system-on-chip using network communication for parallel coarse-grain data processings [dissertation]. Malaysia: Universiti Tunku Abdul Rahman; 2014

[12] Marculescu R, Ogras UY, Peh LS, Jerger NE, Hoskote Y. Outstanding research problems in NoC design: System, microarchitecture, and circuit perspectives. IEEE Transaction on Computer-Aided Design of Integrated Circuit and Systems. 2009. pp. 3-21

[13] Moadeli M, Maji P, Vanderbauwhede W. Quarc: A high-efficiency network on-chip architecture. In: International Conference on Advanced Information Networking and Applications. 2009. pp. 98-105

[14] Chatha KS, Srinivasan K. Layout aware design of mesh based NoC architectures. In: National Science Foundation and Consortium for Embedded System. 2006. pp. 136-141

[15] Becker DU. Efficient microarchitecture for network-on-chip routers. In: Partial Fulfillment of the Requirements for the Degree of Doctor of Philosophy. 2012

[16] Halim DK, Tang CM, Ng MS, Hartono D. Software-based turbo decoder implementation on low power multi-processor system-on-chip for Internet of Things. In: Proceedings of 2017 4th International Conference on New Media Studies (CONMEDIA); 8-10 November 2017. Yogyakarta: IEEE; 2018. pp. 137-141

[17] Wicaksana A, Tang CM, Ng MS. A scalable and configurable Multiprocessor System-on-Chip (MPSoC) virtual platform for hardware and software co-design and co-verification. In: 2015 3rd International Conference on New Media (CONMEDIA); 25-27 November 2015. Tangerang: IEEE. p. 2016

Bio-Inspired Solutions and Its Impact on Real-World Problems: A Network on Chip (NoC) Perspective

Muhammad Athar Javed Sethi, Momil Ijaz,
Huma Urooj and Fawnizu Azmadi Hussin

Additional information is available at the end of the chapter

http://dx.doi.org/10.5772/intechopen.83229

Abstract

Bio-inspired solutions are used to solve the real-world problems as they are able to resolve the complex issues. Already existing bio-inspired solutions are reviewed in this chapter which solved the complex engineering problems. Moreover, this chapter also discusses the implementation of biological brain mechanism in Network on Chip to address the fault-tolerant issues. Network on Chip (NoC) is a communication framework for System on Chip (SoC). Due to routers and interconnect failure, NoC suffers from faults. Therefore, bio-inspired solutions help us to recover from these faults. The techniques from the biological brain were implemented in NoC as the brain is fault tolerant and highly adaptive. Results showed that bio-inspired techniques are performing well compared to the traditional fault-tolerant algorithms.

Keywords: bio-inspired solutions, Systems on Chip, Network on Chip, fault-tolerant algorithms, synapse, neuron

1. Introduction

It is more than 20 years now that scientists and researchers are taking inspiration from biological solutions to build dynamic, real-time and robust applications and systems. The characteristics of adaptability, robustness and resilience to handle failure make biological solutions ideal for solving the real-world complex engineering problems [1]. Based on the literature, the bio-inspired research is basically divided into three main components, that is, bio-inspired computation, bio-inspired systems and bio-inspired networking. In bio-inspired computation, those algorithms are considered which focused on optimization and efficient computing;

however, distributive systems are studied under bio-inspired systems. Bio-inspired networking deals with scalable networking techniques for autonomic organizations and systems. A biological concept should be similar to the real-world problem in order to implement it. It should be realistic to implement it, and the methodology should be clearly defined how to convert it to the real-world system [2].

Biological brain is highly robust and fault tolerant. Synaptogenesis, sprouting, dynamic brain, synapse rearrangement, the long-term potential (LTP) and long-term depression (LTD), nerve growth factor (NGF) and redundant neuron are few fault-tolerant techniques of the biological brain. Synaptogenesis and sprouting algorithms are implemented in Network on Chip (NoC) to make it fault tolerant. **Figure 1** shows the basic structure of 4×4 NoC having mesh topology. NoC is scalable and can be connected in various topologies, which include torus, star, ring, honeycomb and tree. The figure shows NoC having 16 routers and PEs; moreover, PEs can be homogeneous or heterogeneous resources. PEs can be a processor (P), memory (M), cache (C), reconfigurable block (re), digital signal processing (DSP) core or any other intellectual property (IP) cores as shown in **Figure 2**.

There are four broad categories of NoC fault-tolerant techniques, which are deterministic, stochastic and fully adaptive and partial adaptive routing algorithms [3, 4]. Deterministic routing algorithm routes the flit (flow digits; NoC messages are divided into packets, while packets contain flits) in a particular fix direction. XY and YX are types of deterministic routing algorithms. In XY and YX, flits are transferred in the horizontal direction (x-axis) towards destination, PE; later the flits traverse the vertical direction (y-axis), so that flits would reach

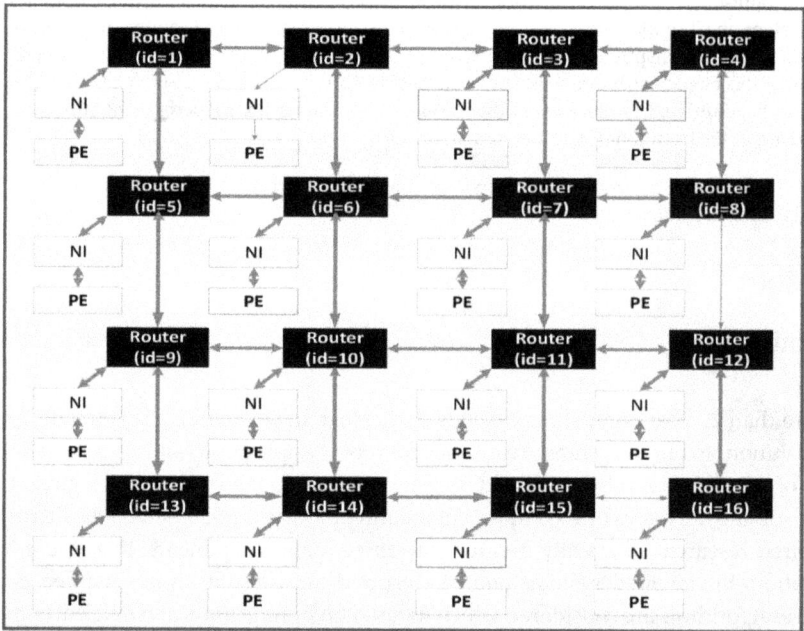

Figure 1. Network on Chip (NoC).

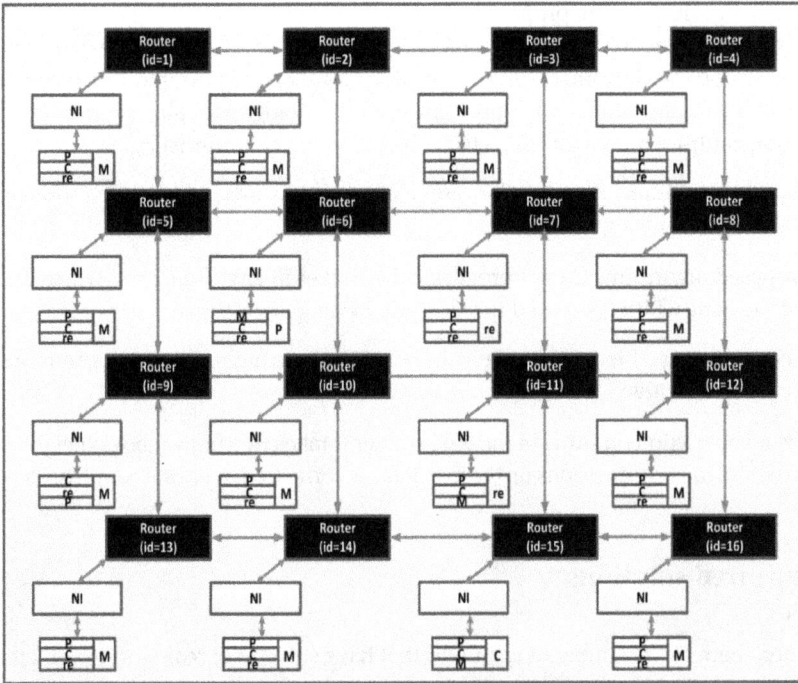

Figure 2. NoC with heterogeneous resources.

the destination PE. The same is the case for YX routing algorithm, in which, flits first cover the vertical direction towards destination PE and later flit traverses the horizontal direction to reach destination PE. Deterministic routing algorithms are not fault tolerant and do not contain adaptivity.

Similarly, stochastic routing algorithm achieves fault tolerance by broadcasting the flits in all directions, until flits reach the destination PE. Stochastic routing algorithms utilize more bandwidth and power. Fully adaptive routing algorithms are adaptive and fault tolerant. However, updating the routing table consumes high power, bandwidth and time. Partial adaptive routing algorithms have restrictions on certain turns in NoC although they are fault tolerant and consume less power. Odd-even is a famous example of the partial adaptive routing algorithm. Odd-even puts a restriction that there will be no flit turn from east to north and from east to south at even NoC columns; moreover, at odd NoC column, no flit turn is allowed from south to west and north to west direction.

Majority of the traditional fault-tolerant algorithms do not completely address the faults in the NoC and have the drawbacks of high packet network latency (time taken from source PE to destination PE) and inter-flit arrival time (time taken between two flits at destination PE), low bandwidth utilization and less throughput. This makes the NoC communication unreliable, and the NoC architecture is not fault tolerant. To overcome the drawbacks of deterministic, stochastic and fully adaptive and partially adaptive routing algorithms, novel biological-inspired fault-tolerant algorithms were proposed. Bio-inspired algorithms are able to address the limitations of fault-tolerant algorithms by mimicking the fault-tolerant and robust mechanism of the brain based on the algorithms called 'synaptogenesis' and 'sprouting'. Novel bio-inspired NoC fault-tolerant algorithms have been inspired by biological brain algorithms. The bio-inspired NoC fault-tolerant algorithms have the following characteristics:

- Bio-inspired algorithms do not follow any fixed path towards a destination compared to a deterministic routing algorithm.

- The proposed algorithms do not broadcast the packet in any directions, which efficiently utilized the bandwidth compared to stochastic routing algorithms.

- There is no routing table in these algorithms, yet they are highly adaptive and robust compared to fully adaptive algorithms.

- There is no restriction on turns in the NoC; rather it takes two hops (nodes) neighbor information before making decisions on the turn it takes compared to partial adaptive algorithms.

2. Bio-inspired solutions

In literature, there are a number of examples that have solved the real-world problems using biological inspiration.

2.1. Ant colony optimization

Ant colony optimization (ACO) algorithm is basically based on the foraging behaviour of the ants. ACO is adopted in the organization of parallel distributed systems. In the biological world, ants collaborate with each other with the releasing special liquid (pheromone) when they find the food. Later, ants group together and follow this shortest path constructed because of the release of this liquid [5].

2.2. Artificial immune system

Artificial immune system (AIS) is inspired by the biological immune system. AIS is used to detect the environmental changes from the normal system and memorize these characteristics automatically [6].

2.3. Autonomic Network on Chip (NoC) using the biological immune system

The robustness and dynamic nature of biological solutions is being used in the NoC. The self-configuration, self-healing and self-optimization are the few dynamic natural characteristics of the immune system used in the NoC to make it autonomic. The bio-inspired NoC (BNoC) behaves and reacts like an immune system as it detects the pathogens (application behaviour and system state change), which enters in the body (system) and delivers a response to heal (adapts to change) it. In BNoC, self-configuration, self-healing and self-optimization are performed in the application, communication and architecture layers. BNoC could react like an immune system against pathogens that have entered the body. It detects the infection (application behaviour or system state changes) and delivers a response to eliminate it (i.e. it adapts to changes) [7].

2.4. Fault-tolerant NoC using biological brain techniques

A neuron is the basic building unit of the biological brain. Neuron rearranges itself in case of damage or due to the broken synapse. Synaptogenesis and sprouting are biological brain techniques being adopted in Network on Chip to make it fault tolerant in case of a router or interconnect failure [8–10].

2.5. Epidemic spreading

The dissemination of information in wireless ad hoc networks is inspired by the epidemic spreading technique. The distribution of information particles or the spread of viruses on the web are linked with the dissemination of information [11].

3. Bio-inspired network on Chip

Bio-inspired NoC fault-tolerant algorithms gain inspiration from biological brain fault-tolerant algorithms to make NoC communication reliable. The biological brain is a complex organ with a network of neurons interconnected by synapses. It works faster than any supercomputer. The neuron is the basic building unit of the brain. There are around 80–120 billion neurons in the human brain. **Figure 3** shows the biological neuron [12]. Two biological brain algorithms, namely, synaptogenesis and sprouting, are adopted and implemented to make NoC fault tolerant. These are self-adapting and self-healing mechanisms. These algorithms help the brain to recover when neurons and synapses get damaged. These bio-inspired NoC fault-tolerant algorithms are implemented using two connection setups or communications services. The two communication services are guaranteed-throughput (GT) and best-effort (BE) services.

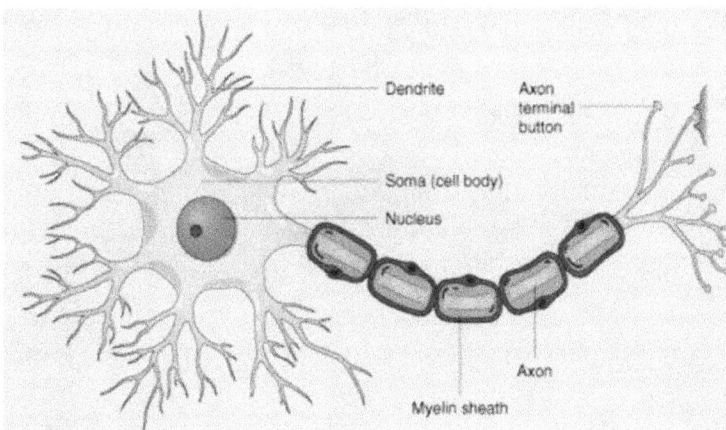

Figure 3. Biological neuron.

3.1. Synaptogenesis

Synaptogenesis is a self-adapting mechanism in the human brain where two neurons attempt to connect and communicate with each other. In this phenomenon, the growth cone (having lamellipodium and filopodia) at the top of the axon and dendrite terminals finds a path to a target neuron. The filopodia actually find the path for a connection with the target neuron. The chemical attractant is released by the target neuron to attract the growth cone. A synapse is formed between the source and target neuron using this method [13]. The synaptogenesis process is shown in **Figure 4**.

3.2. Sprouting

Sprouting is the self-healing mechanism of the brain. Neurons get damage because of hypertension and other different reasons. A new synapse or sprout is created from the already connected source neuron (using chemical adhesion material (CAM)) by establishing a new connection with the target neuron. With the help of this sprout is connected with the target neuron as shown in **Figure 5**. After various experiments on fishes and frogs (amphibians), it is concluded that this feature is not present in humans [13].

3.3. Synaptogenesis-based NoC

The best path is established between the source and destination PE with the help of Synaptogenesis algorithm. Using the algorithm faulty interconnects are detected, and new synapse originates from the neighbor router, which makes the communication reliable and fault tolerant. A credit method is adopted in NoC to detect the faults. Three ns are required to detect the fault as the router does not receive the credit packet from the neighbor router. The concept of credit is shown in **Figure 6**. The flow of flits between routers and PEs is managed by incrementing and decrementing a credit counter. When a port sends a flit, it decrements the credit counter. This specifies

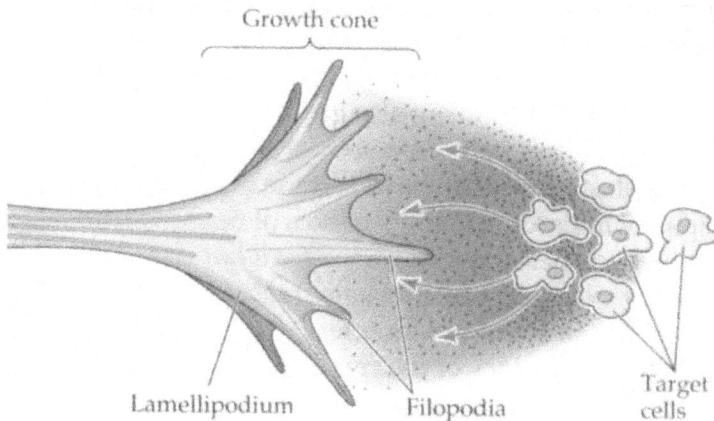

Figure 4. Synaptogenesis.

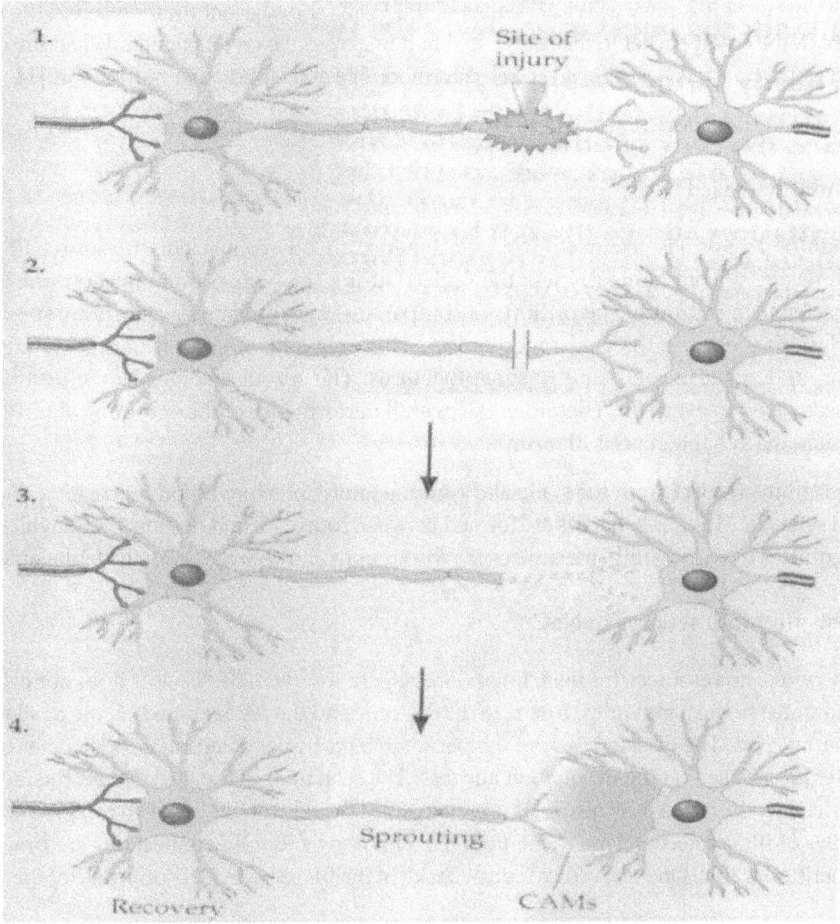

Figure 5. Sprout.

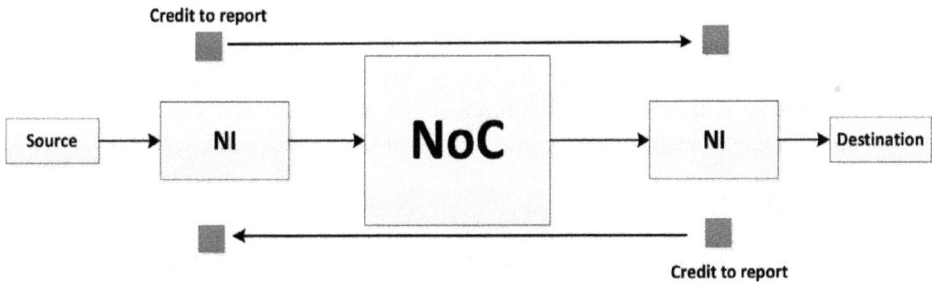

Figure 6. Credit mechanism.

that the output port is busy as it just sends the flit. When the adjacent router receives the flit, it sends back the credit packet to the router so that it increases the credit counter. When the credit counter of the port reaches zero, it cannot send more flits at the particular output port [14, 15].

To improve the fault detection and performance of the NoC, a sprouting algorithm was proposed.

3.4. Sprouting-based NoC

A sprout (or synapse) is originated from the synapse already connected with source PE during the recovery from faults (bypass the fault). During the NoC communication, the faults are detected using sprouting algorithm. As compared to synaptogenesis algorithm, sprouting algorithm does not use the credit mechanism. This helps the algorithm to quickly create a work-around synapse compared to synaptogenesis. The sprout emerges as the destination address is saved in every flit. Therefore, the overall performance of the sprouting algorithm is better than the synaptogenesis algorithm.

The algorithms recover from the static and runtime faulty interconnection by creating a work-around synapse. Multiple synapses are formed between source PE and destination PE which efficiently utilizes the bandwidth, maximizes the throughput and recovers from the faults quickly.

3.5. Best effort (BE) services in NoC

In BE services, no resources (routers, interconnects) are reserved. BE services does not provide any guarantee of the bandwidth. Any source (PE) can send the packet based on the availability of an output port. The decision to send the packet in a particular direction is made on a router-to-router basis based on the destination address. The scheduler of the port detects the fault. In BE services, credit-based flow control is used to control the flow of flits between routers and PEs. Examples of BE connections are cache updates. This shows that BE connections are preferred for noncritical traffic. One of the major drawbacks of the BE services is its unpredictability [16].

3.6. Guaranteed throughput (GT) services in NoC

In GT services, resources are reserved for a particular time period between specific source and destination pair. In GT services, the bandwidth is reserved for particular connections between source and destination pairs. GT connections may underutilize network resources at certain times. This makes the GT connections an expensive option. At certain times, routers and PEs send a burst of data on these GT connections, and then it remains silent for a certain period of time. This leads to the underutilization of network resources. That is why BE services complement the GT services by utilizing the unused bandwidth. Video processing is an example of a GT connection. This implies that GT connections are usually preferred for real-time critical traffic applications [17].

3.7. BE + GT services in NoC

To efficiently utilize the bandwidth of NoC, BE services are provided along with GT services [18]. The flits are saved in best effort queue (BQ) or guaranteed throughput queue (GQ) depending on the source of the flit. Round-robin slot mechanism is used in BE + GT services, whereas GT flits have the highest priority compared to BE flits [19, 20]. The port architecture for BE + GT is shown in **Figure 7**.

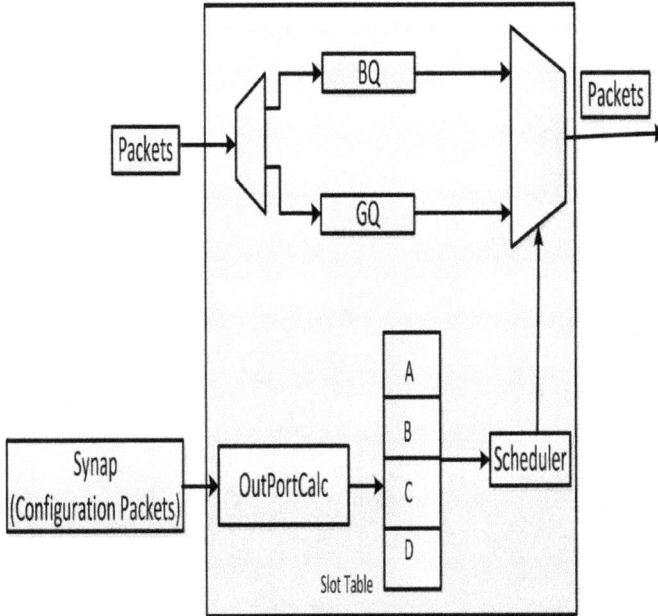

Figure 7. Port architecture.

3.8. Bio-inspired algorithm implementation

The synapse packet is initiated from the source PE and traverses towards the destination using synaptogenesis algorithm. The path is saved in every flit after the creation of synapse to make NoC fault tolerant. During the establishment of synapse, if there is no fault, then flits will traverse on the path. However, if during the communication fault occurs, then neighbor router scheduler detects the fault. The port architecture is shown in **Figure 8**. A new sprout (synapse) emerges from the neighbor router 'inPort' to bypass the faulty interconnect. The port of the router connected with that particular faulty interconnect is blocked through 'Scheduler'. The flits are received at the 'inPort', while the Scheduler is used to send the flit to the neighbor router. The 'OutPortCalc' is used to find the best path for synapse connection.

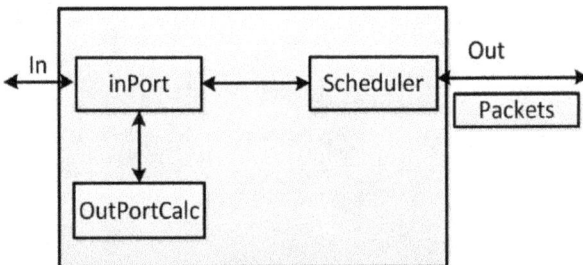

Figure 8. NoC port architecture.

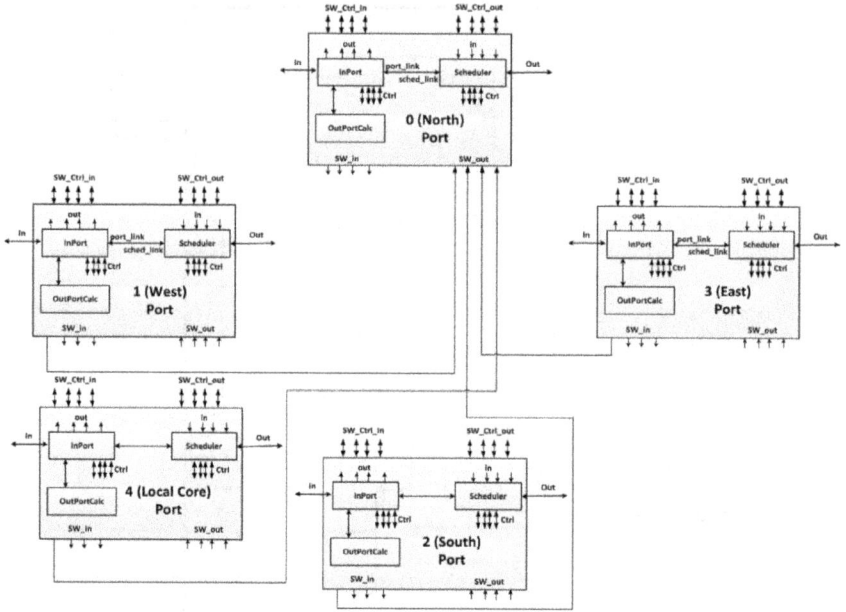

Figure 9. Port path for data flits.

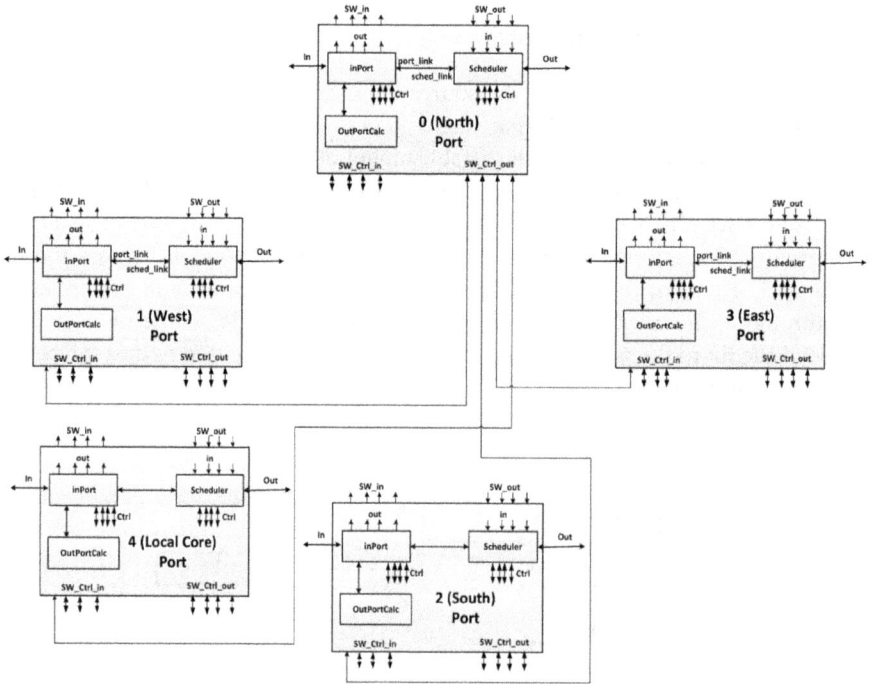

Figure 10. Port path for control flits.

Inside the router, the five ports are connected together using crossbar interconnection. There are separate paths for data and control flits between the ports of the router as shown in **Figures 9** and **10**, respectively. The 'SW_in', 'SW_out', 'SW_Ctrl_in' and 'SW_Ctrl_out' are connected together between different ports of the router. In **Figure 9**, four pins of 'SW_out' of port 0 (north) are connected with the one 'SW_in' pin of port 1 (west), port 2 (south), port 3 (East) and port 4 (local core), respectively. Similarly, four pins of 'SW_Ctrl_out' of port 0 are connected with each 'SW_Ctrl_in' of ports 1, 2, 3, and 4, respectively as shown in **Figure 10**. The same mechanism is used for other ports of a router. The data flits traverse from 'SW_in' of one router port to 'SW_out' of the other port of the same router, whereas control packets traverse on the 'SW_Ctrl_in' and 'SW_Ctrl_out' [21].

With the help of these two bio-inspired algorithms, the NoC becomes robust and fault tolerant and is able to efficiently utilize the bandwidth and maximize the throughput and has less end-to-end latency and inter-flit arrival time than the literature techniques.

4. Results and discussions

In **Figures 11–14**, the bio-inspired algorithm having BE + GT services is compared with literature techniques of XY and odd-even algorithms. As shown in **Figure 11**, the runtime throughput (bits per second) of the bio-inspired algorithm is better than the XY and odd-even by 66 and 58%, respectively.

Moreover, the bandwidth utilization (megabyte per second) by the bio-inspired algorithm is maximum compared to XY and odd-even by 45 and 39%, respectively, as shown in **Figure 12**. Similarly, packet (having flits) network latency (in nanoseconds) of bio-inspired algorithm is less than XY and odd-even as shown in **Figure 13**. **Figure 14** shows that the interflit arrival time of XY and odd-even is more than BE + GT-based bio-inspired algorithm by 49 and 62%, respectively.

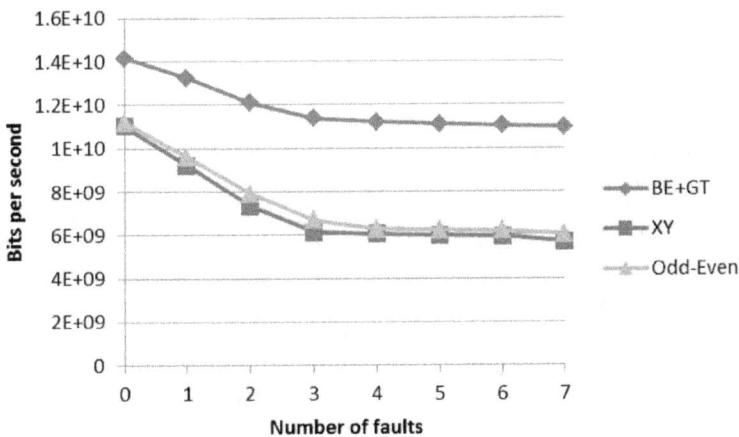

Figure 11. Runtime throughput vs. number of faults.

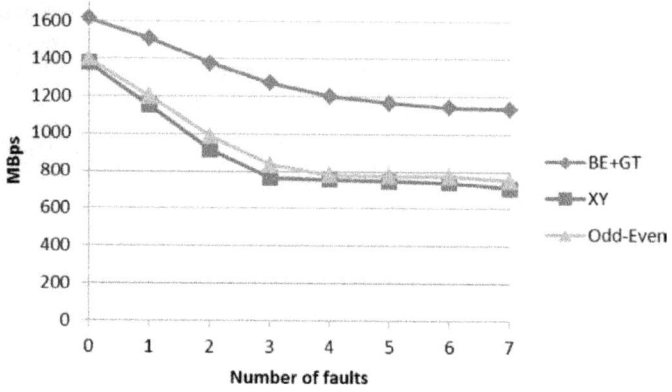

Figure 12. NoC bandwidth vs. number of faults.

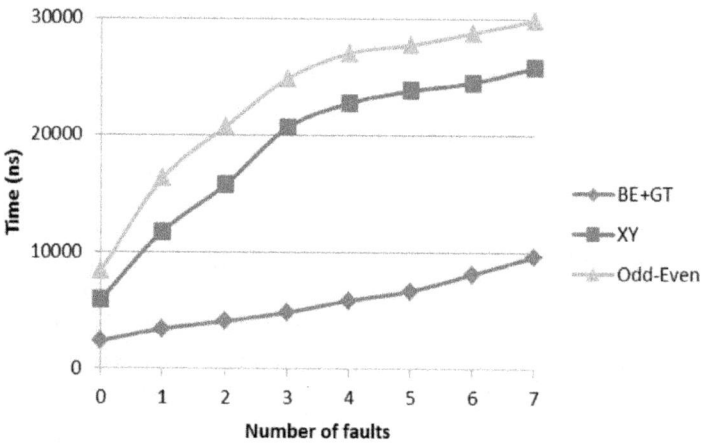

Figure 13. Packet network latency vs. number of faults.

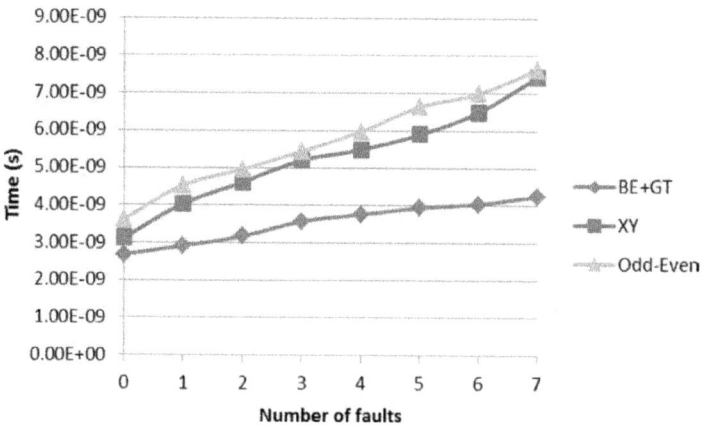

Figure 14. Interflit arrival time vs. number of faults.

5. Conclusion

In this chapter, various biological-inspired solutions are discussed. It is quite wonderful to see how natural phenomenon is able to solve the complex real-world problems. Different biological brain fault-tolerant techniques are presented, which are implemented in NoC to make it fault tolerant. The bio-inspired techniques having BE + GT services are performing better than the literature techniques.

Author details

Muhammad Athar Javed Sethi[1]*, Momil Ijaz[1], Huma Urooj[1] and Fawnizu Azmadi Hussin[2]

*Address all correspondence to: atharsethi@uetpeshawar.edu.pk

1 University of Engineering and Technology, Peshawar, Pakistan

2 Universiti Teknologi PETRONAS, Malaysia

References

[1] Dressler F, Akan OB. A survey on bio-inspired networking. Computer Networks. 2010; **54**:881-900

[2] Dressler F, Akan OB. Bio-inspired networking: From theory to practice. IEEE Communications Magazine. 2010;**48**:176-183

[3] Sethi MAJ, Hussin FA, Hamid NH. Survey of network on chip architectures. Science International (Lahore). 2015;**5**:4133-4144

[4] Athar Javed Sethi M, Azmadi Hussin F, Hisham Hamid N. Review of network on chip architectures. Recent Advances in Electrical & Electronic Engineering (Formerly Recent Patents on Electrical & Electronic Engineering). 2017;**10**:4-29

[5] Drigo M. The ant system: Optimization by a colony of cooperating agents. IEEE Transactions on Systems, Man, and Cybernetics-Part B. 1996;**26**:1-13

[6] Castro LN, De Castro LN, Timmis J. Artificial Immune Systems: A New Computational Intelligence Approach. London: Springer Science & Business Media; 2002

[7] Bakhouya M. Towards a bio-inspired architecture for autonomic network-on-chip. In: 2010 International Conference on High Performance Computing and Simulation (HPCS). USA: IEEE; 2010. pp. 491-497

[8] Sethi MAJ, Hussin FA, Hamid NH. Synaptogenesis based bio-inspired NoC fault tolerant interconnect. In: 2013 IEEE International Conference on Control System, Computing and Engineering (ICCSCE). USA: IEEE; 2013. pp. 46-51

[9] Sethi MAJ, Hussin FA, Hamid NH. Implementation of biological sprouting algorithm for NoC fault tolerance. In: 2013 IEEE International Conference on Circuits and Systems (ICCAS). USA: IEEE; 2013. pp. 39-44

[10] Sethi MAJ, Hussin FA, Hamid NH. Bio-inspired NoC fault tolerant techniques. In: 2014 5th International Conference on Intelligent and Advanced Systems (ICIAS). USA: IEEE; 2014. pp. 1-6

[11] Vogels W, Van Renesse R, Birman K. The power of epidemics: Robust communication for large-scale distributed systems. ACM SIGCOMM Computer Communication Review. 2003;**33**:131-135

[12] Hashmi A, Berry H, Temam O, Lipasti M. Automatic abstraction and fault tolerance in cortical microachitectures. ACM SIGARCH Computer Architecture News. 2011:1-10

[13] Rosenzweig MR, Breedlove SM, Leiman AL. Biological Psychology: An Introduction to Behavioral, Cognitive, and Clinical Neuroscience. USA: Sinauer Associates; 2002

[14] Hansson A, Goossens K, Rădulescu A. Avoiding message-dependent deadlock in network-based systems on chip. VLSI Design. 2007;**2007**

[15] Sethi MAJ, Hussin FA, Hamid NH. Implementation and analysis of biological synaptogenesis technique on nodes and interconnects for NoC fault tolerance. Maxwell Scientific Publication Corp. Submitted: September. 2016;**2015**:483-489

[16] Sethi MAJ, Hussin FA, Hamid NH. Bio-inspired fault tolerant network on chip. Integration, the VLSI Journal. 2017;**58**:155-166

[17] Sethi MAJ, Hussin FA, Hamid NH. Biologically inspired network on chip fault tolerant algorithm using time division multiplexing. In: 2016 6th International Conference on Intelligent and Advanced Systems (ICIAS). USA: IEEE; 2016. pp. 1-6

[18] Rijpkema E, Goossens K, Radulescu A, Dielissen J, van Meerbergen J, Wielage P, et al. Trade-offs in the design of a router with both guaranteed and best-effort services for networks on chip. IEEE Proceedings-Computers and Digital Techniques. 2003;**150**:294

[19] Sethi MAJ, Hussin FA, Hamid NH. Bio-inspired NoC fault tolerant techniques using guaranteed throughput and best effort services. Integration, the VLSI Journal. 2016;**54**:65-96

[20] Sethi MAJ, Hussin FA, Hamid NH. Bio-inspired network on chip having both guaranteed throughput and best effort services using fault-tolerant algorithm. IEEJ Transactions on Electrical and Electronic Engineering. 2018;**13**(8):1153-1162

[21] Ben-Itzhak Y, Zahavi E, Cidon I, Kolodny A. HNOCS: Modular open-source simulator for heterogeneous NoCs. In: 2012 International Conference on Embedded Computer Systems (SAMOS). USA: IEEE; 2012. pp. 51-57

ASIC Testing Methods

Chapter 5

Application of Optical Methods to Electronic Component Stress Analysis

Caterina Casavola, Luciano Lamberti,
Vincenzo Moramarco, Giovanni Pappalettera and
Carmine Pappalettere

Additional information is available at the end of the chapter

http://dx.doi.org/10.5772/intechopen.82714

Abstract

Increasing electronic component reliability is, nowadays, a hot topic both in most advanced applications as well as in electronic devices of common use in everyday life. In fact, requirements in terms of miniaturization of electronics components introduce issues connected with heat dissipation management. Materials, packaging, heat dissipator, and even positioning of the component on the board should be optimized in order to reduce thermal stresses generated in the components, which are one of the most important failure mechanisms of electronics. Thermal stress evaluation is, however, a difficult task due to the size of the elements under testing and to the necessity of measuring small amount of strains. At the same time, any contact with the object under measurement should be avoided not to alter heat capacity of the component itself. In this work, some results of experimental stress analysis gathered using electronic speckle pattern interferometry will be described; it will be pointed out how this approach allows to put in evidence inhomogeneous stress fields undergone by the electronic components and how it is possible to highlight the presence of bad functioning and defects.

Keywords: electronic component reliability, electronic speckle pattern interferometry, bad thermal contact, deformation field, damage detection

1. Introduction

Electronic components are, nowadays, widely spread at any level from simple personal devices to very demanding equipment for aerospace applications. Independently of their specificity, it is a general rule that they are subjected to thermal loading as a consequence of the Joule effect connected to the running electric current. This introduces a mechanical problem because thermal deformations can determine crack initiation and propagation [1, 2]. In this view, the possibility to have accurate and detailed information on the mechanical behavior of the component during working is important. However, due to the complexity of electronic components [3], experimental methods are strictly necessary. Different experimental approaches have been used to the scope such as acoustic microscopy, X-ray, and thermography [4, 5]. Among these, X-ray and infrared thermography are the most common ones. The use of optical methods has also been explored by several authors [6–9]. These methods rely on the modulation in the reflected/diffused light wavefront connected with deformation of the surface. Demodulation of the information allows us to determine displacement field. Interestingly, this can be done without any contact with the component. From the above-mentioned survey, it appears that speckle and moiré are the OT most commonly used in the analysis of electronic chips. Speckle and moiré methods, in particular, have been applied several times in the analysis of electronic chips. In particular, the phase shifting electronic speckle pattern interferometry (PS-ESPI) is very appealing because it does not require application of a grating and can guarantee very high sensitivity [10]. As it was previously underlined, localized mechanical damage can lead to stress/strain concentration that can cause component rupture. If cyclic loading is considered, this mechanism can also be driven by progressive damage accumulation. Moreover, it should be taken into account that the presence of a damage can alter the electrical resistance. This also alters thermal distribution. Proper mounting of the components is a determinant stage in thermal management. In fact, inappropriate mounting can introduce bad thermal contact, which alters thermal dissipation capability of the chip. In this chapter, the PSESPI technique will be applied to analyze displacement fields of components under different conditions. It will be applied on a Darlington transistor (NZT605) to evaluate its mechanical response when powered. The same will also be studied under anomalous loading conditions to understand which kind of effects this introduces on the mechanical behavior. Furthermore, the components will be analyzed when subjected to successive powering cycles in order to understand if deformation of the component is completely recovered at the end of the cycle or some plastic deformations remain. Finally, a voltage stabilizer will be studied; in this case, two different conditions of thermal contact will be analyzed to understand if it is possible, by analyzing the displacement fields, to distinguish between those two conditions. It should finally underline that, in addition to the specific case studies and applications that will be discussed in this chapter, authors intend to put in evidence a more general aspect, which is overall connected to the potentiality and the versatility of an approach based on optical methods. As there is no contact with the measured specimen, optical methods can be used even if the component is highly miniaturized without altering heat dissipation of the component itself. In addition, this can be done independently of the complexity of the analyzed component that can be very high as for the case of ASIC and FPGA components. Finally, it should be pointed out that spatial resolution of this technique is basically connected with resolution of the employed camera. Nowadays, technological developments allow to easily get high resolution camera at low price, so that it is easy to reach micrometric resolution. Hence, dedicate analysis in a critical region or in correspondence of a pad to detect malfunctioning, bad thermal contact, and hot point can be done at a resolution not achievable by thermographic methods. Feasibility of use of the proposed approach to the analysis of other types of chips (e.g., ASIC or FPGA with ball grid arrays) is briefly discussed at the end of the chapter.

2. Materials and methods

The system that was built up to measure the displacement field on the Darlington transistor and the positive voltage stabilizer following the general plan reported in the Introduction section is based on the electronic speckle pattern interferometry. In this specific optical technique, the information is carried out by the random light distribution which is obtained whenever a coherent beam of light illuminates a rough surface. If the illuminated surface is deformed, speckle pattern also changes as a consequence. Here, a double illumination interferometer [11] was built up, which is sensitive to the in-plane horizontal displacement component. **Figure 1** shows the schematic of the experimental set-up utilized in this research.

The setup included a 17 mW He-Ne laser beam (λ = 632.8 nm), which goes through a spatial filter, and it is collimated before illuminating the surface. An intensity beam splitter is used to divide the beam into two paths. The emerging beams are reflected by mirrors M2 and M3 and then directed at the same angle toward the surface. In this way, symmetric double illumination is obtained. A black and white Marlin FireWire Allied Vision Technologies camera was used to capture images whose sensor has 1628 × 1236 pixel. Camera was equipped with imaging lenses with field-of-view = 21 mm. The optical axis of the camera is perpendicular to the component under analysis. Temporal phase shifting approach [13] was utilized in order to recover phase of the speckle correlation fringes. For that purpose, a piezoelectric translator is inserted into one of the arms of the interferometer in order to allow phase shifting. In particular, the five-step technique was adopted in which subsequent $\pi/2$ shifts are introduced, so that the following sequence is generated: φ_n = 0, $\pi/2$, π, $3\pi/2$, 2π (n = 1,…, 5). It should be noted that the phase of the first and last image is the same. The following expression can be used to obtain the phase difference encoded in the speckle correlation fringes:

$$\Delta \varphi(x, y) = \tan^{-1} \frac{2\,I_2(x, y) - I_4(x, y)}{2\,I_3(x, y) - I_5(x, y) - I_1(x, y)} \tag{1}$$

where $I_1(x,y)$, $I_2(x,y)$, $I_3(x,y)$, $I_4(x,y)$ and $I_5(x,y)$ are light intensity values of the correlation fringe patterns generated for each phase shift. In order to improve the data quality, it was decided to preprocess fringes before extracting phase information. After some preliminary investigation, it was decided to apply a median filter to the image. This filter replaces, for each pixel, the median value calculated over the neighborhood pixels. A 7 × 7 pixel window was adopted for these specific measurements. With the same aim of improving data quality, it was decided to apply a white sprayed layer to the inspected components. This was done to improve fringe contrast in order to have benefits in terms of final accuracy. The in-plane displacement components u of each generic point P(x,y) of the specimen can be obtained by the following expression:

$$u(x, y) = \frac{\Delta \varphi(x, y)}{2\pi} \cdot \frac{\lambda}{2 \sin \vartheta} \tag{2}$$

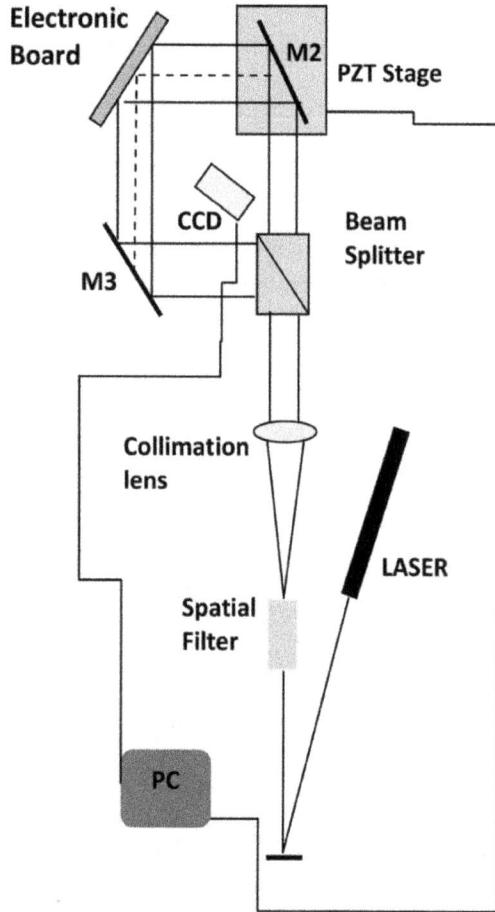

Figure 1. Schematic of the implemented phase shifting interferometric speckle setup [12].

The sensitivity λ/2sinθ of the optical setup is 447 nm for the given optical-geometrical configuration; this means that points belonging to two adjacent fringes experience a relative displacement along the horizontal in-plane direction of about 0.5 µm. First tests were performed on a NZT605 Darlington transistor mounted on a S9004 electronic board adopted in satellite space application. The board is a DC/DC convertor with a dual output ±5.7 V. The indicated transistor was subjected to analysis because it was indicated, by the producer, to be the most critical element in terms of reliability. More in detail, the analyzed component is an NPN Darlington transistor whose gain factor can be up to 10,000 of the input current. It was mounted on a SOT-223 type package (**Figure 2**), which is specifically designed for this kind of application because it guarantees improved performances in terms of thermal dissipation.

The measurement procedure starts by acquiring a set of five reference images, one for each phase step. After this, starting step power was applied to the component, and speckle patterns were acquired according to a given temporal sequence. These speckle patterns are subtracted one by one from the five reference patterns. When this is done, after proper preprocessing of the images as described above, Eq. (1) is used to calculate wrapped phase, while the spanning tree algorithm implemented in the Fringe Application software is used to obtain the unwrapped phase [14]. Finally, Eq. (2) is used to obtain the displacement maps. The experimental plan was organized in order to understand the thermomechanical behavior of the component both in normal working condition and in the presence of a damage, which can cause malfunctioning of the board. After some preliminary attempts and in accordance with the indications of the producer, this last condition was simulated by powering the board with 29 V instead of the nominal powering voltage of 17 V. This corresponds, at the level of the analyzed component, to an applied voltage 50% higher than the maximum suggested voltage so that this condition simulates working under severe conditions. Having this in mind, two types of tests, that is to say static and fatigue tests, were performed. In the static test, voltage is applied to the board at a given instant time t_{0}, and the power is kept on for the following 10 minutes. During this phase, speckle patterns are acquired in order to evaluate progression of the u-displacement field in the heating stage. Ten minutes are enough to reach a stationary state as it was inferred from preliminary investigations. At the end of this stage, the board is powered off, and the following 10 minutes are monitored in order to get information on the evolution of the u-displacement fields during the cooling phase. The same procedure was finally repeated after 90° rotation of the board in order to measure the v-displacement field as well, and in such a way, complete description of the in-plane displacement field is obtained. For analyzing the fatigue behavior of the NZT605 component, cyclic powering loading was considered. Two different loading cycles were implemented, and for each of them, two levels of maximum voltage were considered corresponding, as for the static tests, to nominal and anomalous working conditions. A timer was put in series between the power supply and the analyzed board. Timer allows regulation of the on and off time. Experiments were performed under

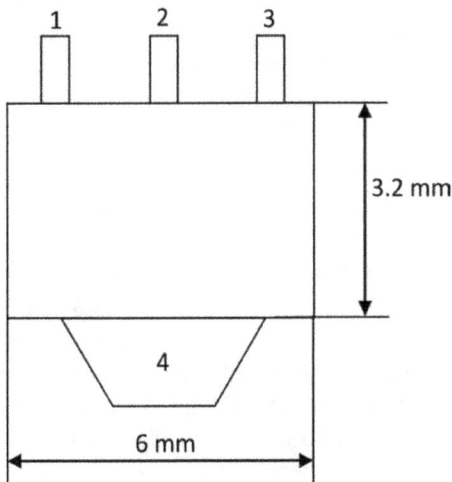

Figure 2. Schematic of the SOT-223 packaging.

slow loading conditions, that is to say, with a time duration of each loading cycle T_{slow} = 400 s and under fast loading conditions, that is to say, with a time duration of each loading cycle T_{fast} = 10 s. Duty cycle is fixed to 50% for both experimental conditions. The duration of the entire fatigue test was 8 h. It was divided into four steps of 2 h each. At the end of each step, the board is powered off for half an hour, and finally a speckle pattern is acquired. This speckle pattern is subtracted by the five reference patterns recorded at the beginning of the test, and displacement maps are obtained as described for the static test. In this way, it is possible to extract the information on the residual deformation of the component as a consequence of the loading cycles. Five different measurements were performed for each of the loading conditions and repeated for two levels of voltage, 17 and 29 V. As it was stated for the static tests, those two levels are representative of the normal and anomalous working conditions. Finally, a second set of experiments was performed on a LM7818 voltage stabilizer. In this case, the main scope of the analysis was to understand if the PSESPI system could capture difference in the deformation field connected with a bad thermal contact. This was simulated by bad screwing of the component to the board so that the lower part of the body of the component does not adhere perfectly to the board. This is a condition that can appear as a consequence of inappropriate mounting procedure or that can arise as a consequence of warping of the board during exercise. Reference pattern collection and fringe analysis follow the procedure described before. Then, the component is powered to the nominal input power, and 5 s is waited before collecting a speckle pattern. Measurements were repeated five times for each condition.

3. Results and discussion

In **Figure 3**, the relative displacement as recorded along the indicated cross section is reported as a function of the test time. In particular (**Figure 3A**), for the case of the *u* horizontal displacements, the top and the bottom cross sections are reported, while for the case of the *v* vertical displacements (**Figure 3B**), the middle and the bottom cross sections are shown. It is clearly visible how, after 10 minutes of board powering, a tree time higher level of *u*-displacement is observed in case of anomalous working conditions than normal working conditions (3.01 μm vs. 2.73 μm). Analogous conclusions can be done by observing the *v*-displacement field.

In fact, when over voltage is applied, a 2.5 times higher displacement in the central section of the component (1.46 μm vs. 0.56 μm) can be observed. Deformation field is not homogenous, and in fact, displacement in the middle of the component is 30% larger than in the bottom part (1.46 μm vs. 1.11 μm). This general consideration holds for both normal working and anomalous working conditions. However, it can be observed that those differences are progressively recovered during the cooling state, and at the end of the test, no final residual displacement can be observed. This different behavior during the heating and the cooling stage is interesting to be observed, and it is a simple consequence of the high complexity of the thermo-mechanical problem. To get more understanding, it should be recalled that the SOT-223 package (**Figure 2**) is properly designed to improve heat dissipation capability especially through the pad 4 where high level of current, up to 1.5 A, can flow. However, especially when over voltage is applied, pad 4 is not able to dissipate completely the heat so

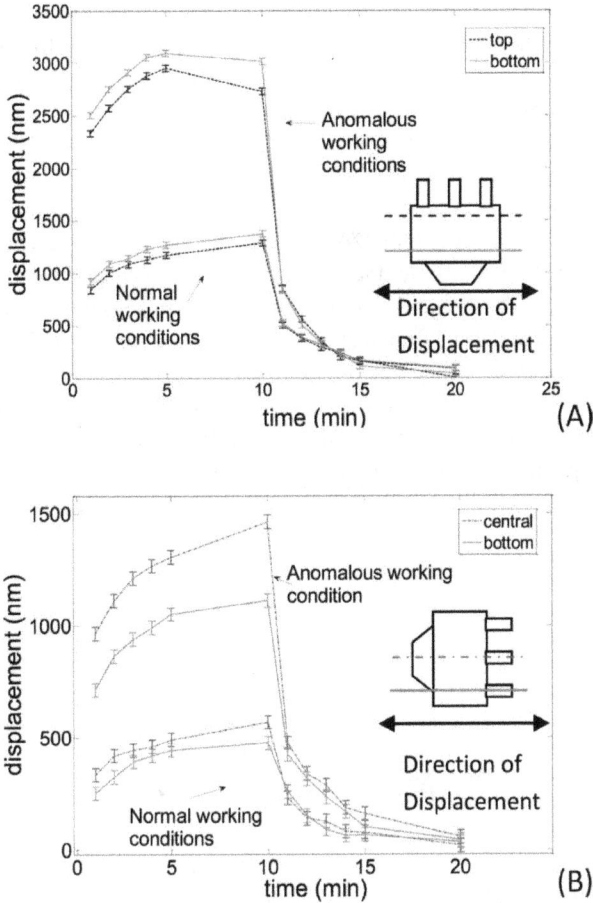

Figure 3. Behavior of the displacements recorded along the component during the static tests. (A) horizontal displacements, (B) vertical displacements.

that thermal gradient along the component is higher and this results in a less homogenous displacement field. After powering off the board, no current flows through the collector so that overall thermomechanical behavior rapidly returns to be uniform. This is an important consideration as improving heat flow and dissipation is a critical issue for high power components that can have impact on their reliability and lifetime. Pad 4 is designed to the scope of improving power dissipation but, nevertheless, our measurements that still strain values up to $500 \ 10^{-6}$ m/m can be observed.

In **Figure 4**, the results in terms of residual deformation are reported for the 8 hour fatigue tests performed in this study. In particular, the reported data are referred to the region close to the bottom of the component that is to say in the nearby of the pad 4 where both u displacements and v displacements are higher. Results for normal working conditions are reported in **Figure 4A**, while results referred to anomalous working conditions are reported in **Figure 4B**.

Statistical dispersion was calculated over five measurements, and it is comparable to that observed for the static tests. It can be observed that if a 10 s cycle duration is considered (i.e., high frequency, fast cycle), no significant residual displacement field can be observed. Final displacement obtained after 8 cycling hours was 50 nm in case of normal working and 200 nm in case of anomalous working that is to say for times higher in this last condition. However, if a 400 s cycle duration is considered (i.e., low frequency, slow cycle), it can be observed that residual u displacements become considerably higher. In fact, a final residual displacement of 202 nm is recorded for 17 V of applied voltage, which increases up to 691 nm when 29 V is applied. In general, by data observation, it can be said that residual deformation accumulates during the test and that it increases about 3.5 times if anomalous working conditions are implemented. If loading histories are compared for the high frequency and low frequency fatigue tests, it should be noted that, in the former case, the board is subjected to a total amount of 2880 cycles, while in the second case, the total number of cycles is 72. When slow cycling is considered, it can be inferred that heat has more time to propagate across the component. Consequently, larger gradients of temperature distribution can build up, and this

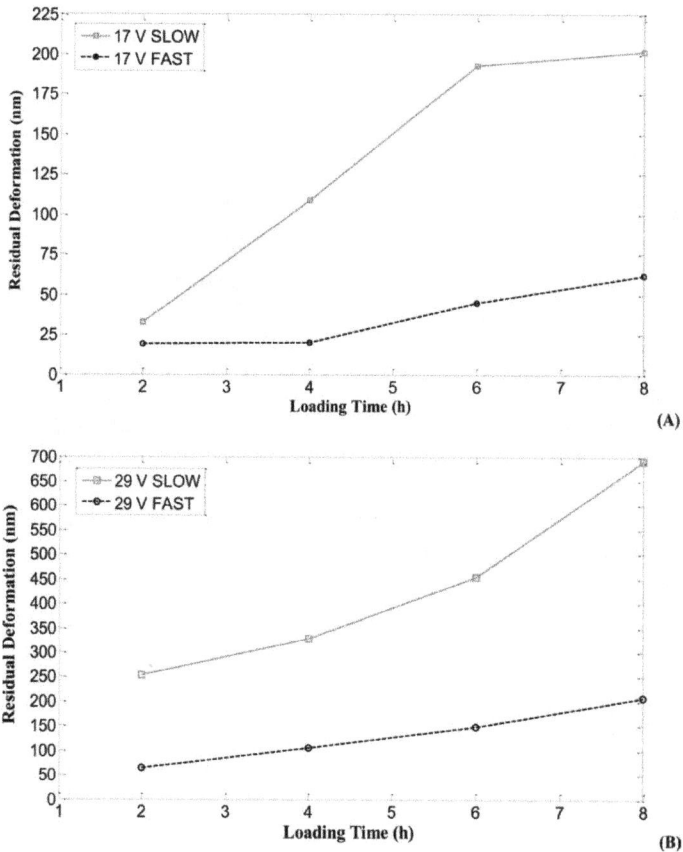

Figure 4. Residual displacements measured at 17 V (A) and 29 V (B) for the cases of slow and fast cycles.

leads to more asymmetry in structural response, which turns, at the end, in larger residual displacements. Of course, the higher the voltage applied to the specimen, more evident is this phenomenon. This leads to another important observation. It should be taken into account that accelerated tests are a powerful tool for reliability assessment [15]; however, the behavior illustrated before shows that proper choice of the voltage must be done in order to have significant findings. Another interesting observation that can be done in relationship with the experimental data is that the increase of residual deformation, when low cycling frequencies are considered, is bigger in anomalous working condition as expected.

Figures 5 and 6 show the speckle correlation fringes recorded after 5 s of powering of the component [16]. In particular, Figure 5 shows correlation fringes recorded in case of good thermal contact, while Figure 6 shows fringe correlation in case of bad thermal contact. It appears soon very evident how fringe pattern changes very much between those two situations. In the case of good thermal contact, it appears evident the inhomogeneous strain distribution connected with the different levels of current flowing in different parts of the components. In the case of bad thermal contact, instead, the displacement field appears to be dominated by the shear component. This can be explained if it is taken into consideration that bad thermal contact configuration was simulated by not perfect adhesion of the lower part of the body of the component to the board. This leads to have two distinct regions in the upper part and in the lower part of the component with very different heat dissipation capability;

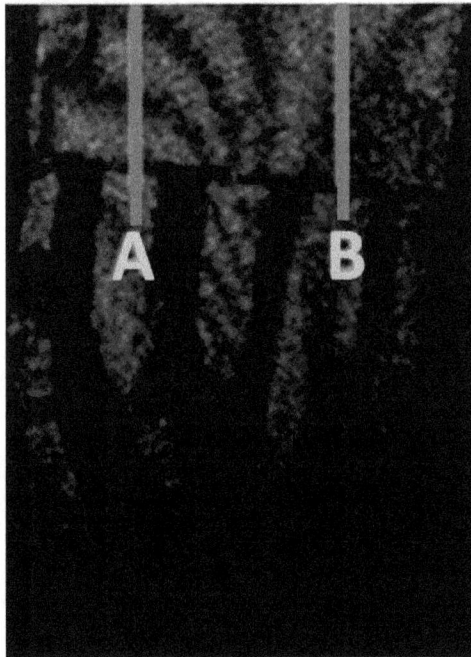

Figure 5. ESPI correlation fringe pattern recorded on the LM7818 component when good thermal contact between component and board is present.

as a consequence, a temperature gradient which follows the vertical direction arises, and this leads to the appearance of dominant shear deformation. Another interesting observation can be done if the left part and the right part of a positive voltage stabilizer are compared. It should be underlined that input current flows starting from the pin in the left side of the component, while output current flows starting from the pin in the right side. Deformation field appears very different in the two sides mainly as a consequence of different levels of input and output current. This is particularly evident when good thermal contact is realized. Conversely, in the case of bad thermal contact, shear component is dominant and tends to mask this effect. In **Table 1**, the shear strain ε_{xy} is reported. In particular, this was calculated by following the two lines A and B reported in **Figures 5** and **6**. Section A and B were taken in correspondence of the input and the output pins. It can be easily argued how measurement of ε_{xy} can be efficiently used to detect the occurrence of bad thermal contact. In this case, in fact,

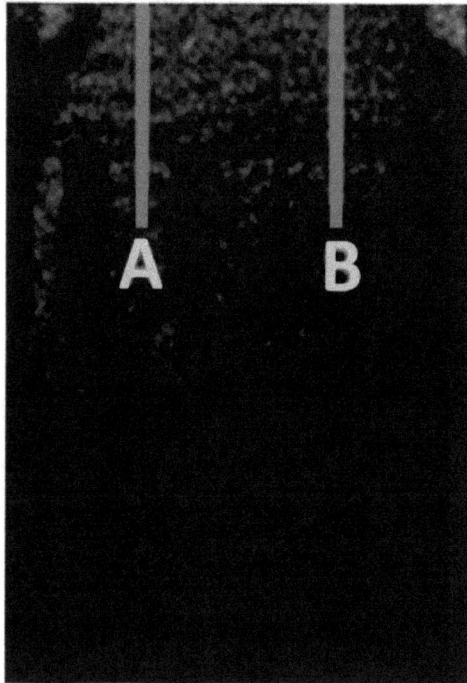

Figure 6. ESPI correlation fringe pattern recorded on the LM7818 component when bad thermal contact between component and board is present.

Good thermal contact		Bad thermal contact	
ε_{xy} (10^{-6} m/m) along A	ε_{xy} (10^{-6} m/m) along B	ε_{xy} (10^{-6} m/m) along A	ε_{xy} (10^{-6} m/m) along B
60	200	850	920

Table 1. Shear strain recorded on the LM7818 component in case of good and bad thermal contact.

$\varepsilon_{xy} = 200 \ 10^{-6}$ m/m was measured in the nearby of the output pin when good thermal contact is present, while this value rises up to $\varepsilon_{xy} = 920 \ 10^{-6}$ m/m in the case of bad thermal contact.

4. Conclusions

In this work, we have reported some results obtained by applying electronic speckle pattern interferometry to measure displacement field of electronic components under different conditions. We have demonstrated how this approach captures the complex deformation field occurring in electronic components. Moreover, the sensitivity of the method is enough to record the small residual deformation field that is accumulated at each powering cycle. This is an important aspect that can have effects in terms of determining reliability of the components under different conditions. Furthermore, it was also found that anomalous conditions as those referred to bad thermal contact can be clearly put in evidence by analyzing the recorded fringe pattern. In this line, this approach can be promising as a diagnostic tool to detect bad mounting conditions. In this specific work, a pixel size of about 10 µm was available, and this was enough to capture details in strain distribution for the analyzed component and to detect, for example, difference in correspondence of different pins. One interesting issue may be how to use ESPI for inspecting thermomechanical behavior of other chips with more complex geometry such as, for example, ASIC, FPGA, and CLPD. This task becomes very challenging especially when the contact between chip and board is realized through ball grid arrays as BGAs include much smaller contact surfaces than the large pad of the SOT-23 package tested in this study. This may entail different selections of sensor type (e.g., to increase spatial resolution) and imaging optics (e.g., to achieve higher magnification). Besides changing properly the field of view and resolution of the sensor, the inherent properties of the speckle field provide immediate information on any irregularities/anomalies in thermal deformation. In particular, local decorrelation or defocusing helps to identify critical regions. Major issues such as, for example, short or open circuits caused in BGAs by thermal stresses can be detected by simply looking at the ESPI pattern structure just by a visual inspection.

In perspective, possibility to miniaturize the system by making use of optical fibers and to use it as a diagnostic tool to test a specific component or a subpart with respect to issues connected with bad contact or damage presence is foreseeable. The large amount of information available from ESPI measurements may certainly help designers to optimize chip geometry and PCB layout. By critically comparing design alternatives, it will be possible to select the best solution that allows risk of failure to be reduced by a great extent.

Author details

Caterina Casavola, Luciano Lamberti, Vincenzo Moramarco, Giovanni Pappalettera and Carmine Pappalettere*

*Address all correspondence to: carmine.pappalettere@poliba.it

Department of Mechanics, Mathematics and Management, Polytechnic of Bari, Bari, Italy

References

[1] Tummala RR. Fundamentals of Microsystems Packaging. New York, USA: Mc Graw-Hill; 2001

[2] Ulrich RK, Brown WD, Ulrich RK. Advanced Electronic Packaging. Chichester, UK: Wiley-IEEE Press; 2006

[3] Jang JW, Suk KL, Paik KW, Lee SB. Measurement and analysis for residual warpage of chip-on-flex (COF) and chip-in-flex (CIF) packages. IEEE Transactions on Components, Packaging and Manufacturing Technology. 2012;2(6111273):834-840

[4] Wolter KJ, Oppermann M, Heuer H, Köhler B, Schubert F, Netzelmann U, et al. Micro- and nano-NDE for micro-electronics. In: Proceedings of the IV Panamerican Conference of NDE. Buenos Aires (Argentina); 2007

[5] Sharpe WN, editor. Handbook of Experimental Solid Mechanics. New York, USA: Springer; 2008

[6] Han B, Guo Y. Determination of an effective coefficient of thermal expansion of electronic packaging components: A whole-field approach. IEEE Transactions on Components, Packaging, and Manufacturing Technology: Part A. 1996;19:240-247

[7] Nassim K, Joannes L, Cornet A, Dilhaire S, Schaub E, Claeys W. High-resolution interferometry and electronic speckle pattern interferometry applied to the thermomechanical study of a MOS power transistor. Microelectronics Journal. 1999;30:1125-1128

[8] Genovese K, Lamberti L, Pappalettere C. A comprehensive ESPI based system for combined measurement of shape and deformation of electronic components. Optics and Lasers in Engineering. 2004;42:543-562

[9] Sciammarella CA, Lamberti L, Pappalettere C, Volpicella G, Sciammarella FM. Measurement of deflections experienced by electronic chips during soldering. Journal of Strain Analysis for Engineering Design. 2006;41:597-608

[10] Cote KJ, Dadkhah MS. Whole field displacement measurement technique using speckle interferometry. In: Proceedings of the 51st Conference on Electronic Components and Technology. Orlando (FL), USA; 2001. pp. 80-84

[11] Leendertz JA. Interferometric displacement measurement on scattering surface utilizing speckle effect. Journal of Physics E: Scientific Instruments. 1970;3:214-218

[12] Casavola C, Lamberti L, Moramarco V, Pappalettera G, Pappalettere C. Experimental analysis of thermo-mechanical behaviour of electronic components with speckle interferometry. Strain. 2013;49:497-506

[13] Ghiglia DC, Pritt MD. Two-Dimensional Phase Unwrapping. Theory, Algorithms and Software. New York, USA: Wiley Interscience; 1998

[14] SmartTech Ltd. Fringe Application 2001 Version 1.0, Warsaw (Poland); 2001. www.smarttech.pl

[15] Shiratori M, Qiang Y. Fatigue-strength prediction of microelectronics solder joints under thermal cyclic loading. IEEE Transactions on Components, Packaging, and Manufacturing Technology: Part A. 1997;**20**:266-273

[16] Casavola C, Pappalettera G. Strain field analysis in electronic components by ESPI: Bad thermal contact and damage evaluation. Journal of Nondestructive Evaluation. 2018;**37**:11

www.ingramcontent.com/pod-product-compliance
Lightning Source LLC
Chambersburg PA
CBHW070241230326
41458CB00100B/5761